THE TEAM PLAYER

How to become a highly effective team player

TOM LAWRENCE

Orders: Please contact: https://highlyeffectiveleader.com, tom@highlyeffectiveleader.com

ISBN: 978-1-8383295-7-0

First published 2024

Printed in Great Britain

DEDICATION

To my Mum & Dad, my family, and my friends.

I love you all and I am so appreciative of all the support you have given me throughout my life.

CONTENTS

Leadership is not about you, but it starts with you. Leadership is influence.

CHAPTER 1
TRUST

Grow yourself, grow your influence

Make a commitment to grow yourself. Do not break that commitment until you leave this planet.

Growing you and focusing on personal growth is without doubt the most important thing a person can do for themselves. If you want to live in this life, and not just exist, then you need to constantly grow…every day. The more you grow, the more freedom you will have and the more options you will create. With more options and freedom, your life can only improve each day.

When you are at work, you are part of a team, and an organisation. So, everyone will grow and benefit when your team creates more options for the organisation. You and your team will grow and benefit when your leader creates more options for the team. It is a win-win situation for all.

The only time it will not be a win-win situation for all is when you, the team, or the organisation do not create any options. If that is the case, then it will be lose-lose for all.

If you would like your team to improve, then you must take the lead on this. You cannot just wait for something to happen and hope that things get better. You need to stand up and lead the team. So, the only way you can do this is by starting with ***YOU!*** You must grow yourself before you start growing your team. As you start to grow, the team will automatically grow with you. But it must start with you.

The key to growing yourself and the team is to be "growth oriented" and not "goal oriented". When you are growth oriented you are striving towards growth, which is not final. Growth is never ending. When you are goal oriented, most people set a goal, achieve it and then stop growing because they don't set another goal. So, it is final, and for most people there is no need to keep growing because the goal is achieved. This is unfortunate because these people have a lot more growth potential and can improve so much more. However, when you are growth oriented, you are constantly setting new goals to strive towards your overall *life mission.* My life mission is to constantly grow and help others to grow too.

To be growth oriented and stay on that path, you must be disciplined. To be disciplined you must be willing to make sacrifices. Making a sacrifice is when you give up something that is valuable to you. For example, time on activities you enjoy, or money spent on material things. But it doesn't stop with sacrifice. Making a sacrifice gives you the opportunity to grow, it is not the cause of your growth. So, when you make the sacrifice, you must take the opportunity for growth that you have just created for yourself. If you take the opportunity, you have just created, this is when you must be disciplined to see the opportunity through, so you can grow. You are investing your time/money in your personal growth, and this really does take discipline. Discipline will transform sacrifice into growth.

A highly effective team player is focussed on their own and their team's personal growth. They want to climb the ladder and reach a higher growth level. A low performing team player is not interested in climbing the ladder or reaching a higher level, they would prefer to coast instead. They would just like to get by in life as personal growth doesn't really have any meaning for them. I have worked with and met lots of people who are like this, and I find it a bit sad that they don't want to explore the potential that they have.

Highly effective team players consider themselves a leader, and they also consider every one of their teammates to be leaders too. No matter what stage of growth their teammates are in, or even if they still need to be developed. So, for a highly effective team player to grow in leadership, they will require leadership development. It is their leadership development that is a focus of theirs, along with their personal growth. They understand that to lead you must have influence. They understand that leadership is influence and to lead you do not need to be in a leadership position, or have a leadership title, or even have the authority of the team. They understand that to grow as a leader, then you must develop your character before you develop your competency.

When I was part of a team before I got my first leadership position and title,

I studied leadership. I read leadership books, I watched leadership videos and attended leadership courses. I did this because I was focussed on my personal growth, I was focussed on helping my team grow, and I was focussed on being a highly effective team player. While I was doing this I was also growing and strengthening my character. When a person starts a new job, they are given all the training required to be able to do the job. Hence, they are developing their competency. It is very rare for an organisation to put their people through training that will help them develop their character. So, we must do it ourselves like I was just describing.

Throughout this book, I am going to use my own personal stories throughout my career in the engineering industry, being part of teams, and leading teams. In September 1999, I left school at 16 years old and became an apprentice mechanical engineer. I have been in engineering for 20 years and I have worked in the automotive, gas/heating, and rail industries. I have also worked in Liverpool (my hometown), Edinburgh, Glasgow, and London in the UK. I have also spent a short time working in India.

This book is not about engineering, it is about people. So, you will be able to relate to my stories, and you may have had similar experiences in your area of work and industry. Or, you may know other people who had similar experiences that you can relate to.

The principles to becoming a highly effective team player can be applied to any role, in any industry in the world. However, I will be using my engineering background as case studies on how I applied these principles in this book.

Either way, you will be able to learn from me. Your role through reading this book is to look for the lessons, and together we can make little discoveries that everyone is capable of sharing and teaching to another person.

When we went to school, college, or university, it was and still is classed at being "educated." However, I beg to differ. In The United States, they use the term "educated" too, but they also use a term "schooled." When we were at school that is exactly what we were being.... schooled.

Education is a lifelong journey, it does not stop when we finish school, college, or university. So, we need to get very serious about our education because for us to grow, we need to be constantly educating ourselves. Luckily for me, I realised this when I was in my early 20's and I was working in Liverpool for their train operating company. That is when I started to read books on leadership and watch leadership videos, and I have been doing so ever since.

This was in 2006 and at that time I was doing a part time engineering degree, but I didn't want to just rely on that. I wanted to take my education further. I knew if I took my education further and constantly learned, then I would be growing, and I would be in a much better position to grow my team. It wasn't easy to get started on this because I had not read a book from beginning to end since I was in school about 10 years earlier. I didn't think I needed to anymore…I was wrong.

So, I made a lifelong commitment to educating myself, to grow, lead myself, and eventually to lead others. Over 10 years later I am still honoring that commitment. What commitment have you, or will you make to yourself?

To improve does not mean we must be bad at something. We can always improve on our current best.

Build your character

Who you "claim" to be is not who you really are. If you don't know who you are, then others will know before you. Find out who you really are.

Leadership is influence and to be a highly effective team player, we need to have influence. So, how do we have, and how do we increase our influence with our team to become a highly effective team player? We need to gain the team's ***TRUST***.

If the team trust us, we can work together as a team. If the team don't trust us, we all become individuals in a team. If the team trust us, we can achieve many things as a team. If the team don't trust us, then we are not going to achieve a lot. If the team trust us, we can grow together. If the team don't trust us, our growth is jeopardised.

Trust is the thread that goes through every team in every industry in the world. That is the most common thing that every great team has in common…trust. This is the thing that turns good teams into great teams. Then from great teams into highly effective teams.

So, what can you do to influence your team trust you? Well, first, you cannot *make* someone trust you. But what you can do is make decisions and carry out acts that make you trustworthy. So, you need to decide whether you want to be trustworthy or not before you can become trustworthy. When I made the commitment to grow every day, I made the decision then to be a trustworthy person. What decision are you going to make?

What you will learn in this book are the principles to become a highly effective team player, and how to help others to do the same. If there is no trust then there is no way you will become a highly effective team player, or a highly effective leader. It doesn't matter what position you have within your team. Trust is the foundation of a highly effective team. To build a strong foundation of trust within your team is a responsibility for every team member. So, you must accept the responsibility right now to help build that trust.

I started my career when I was 16 years old as a mechanical engineering apprentice. It was 1999 and I was working for a global automotive manufacturing company based in Liverpool. They built the assembly lines that would cut and shape car engines, for some of the biggest car companies in the world. For example, Ford, Jaguar, BMW, and Mitsubishi. My job for the 4 years I was there was to help build those assembly lines. I loved the work, but I didn't really care for the teams I was part of. In fact, we were not teams at all, we were individuals.

We all had our own individual station on the assembly line and our job was to basically get through the day. We just had to complete the work that was given to us in the morning by the supervisor. That was it. It suited me fine because in those days I was quite shy and didn't like talking a lot. I just wanted to finish my work and go home. I wasn't interested in building trust with anyone. Or building rapport with any colleagues. If I completed what I had been assigned, that was all the trust I needed.

Later, in my career when I took higher level jobs, I had to change my thinking and how I saw my career.

By this time, I was no longer an apprentice, I was a fully qualified engineer and working in a team of other qualified engineers. I was no longer the least ranked in the team, so I had more responsibilities. I was now working in the railway where teamwork is imperative in the engineering departments. Our role was to fix the trains in quick enough time so they can go back into service in time for the customers. Otherwise, we would have a lot of unhappy people stranded at railway stations. So, to accomplish our tasks, we were going to work together as a team, and we needed to trust each other. I needed to ensure that I was a trustworthy team member, and I had to see my teammates as trustworthy too. Trust is everybody's responsibility, and it is a two-way street.

I needed to get over my shyness, start engaging with people, and build relationships at this stage of my career. If I was going to be successful, then I could no longer be thinking just about myself anymore. I needed to be thinking of others, especially those on my team. So, when I saw an opportunity to build a relationship or make one stronger, I took it. I took opportunities to practice my leadership in the team. By doing this I was becoming more trustworthy, and I started helping others to become trustworthy too. This was naturally building stronger trust throughout the team, and my influence was organically increasing with my teammates.

My career has been 20 years long so far, and I have had a lot of experiences, made many mistakes, and learned from them. I have also learned a lot of lessons through working with and alongside leaders in my industry. Whether they had a leadership position or not, they were still leaders. I have used everything I have learned over the last 20 years to write this book. I want to help you become successful in your career and life. I want to help you become a more valuable, trustworthy person, and a highly effective team player. I want to help you to help others and share what you have learned with them so they can follow in your footsteps.

This is my third book, and the thread through all three of them is that the principles you learn will all make you a more *valuable* and *trustworthy* person. I started reading leadership books in 2006, and everything I have learned and applied has made me a more valuable and trustworthy person. I want to do the same for you, and I want to help you to do the same for others. It doesn't matter what role you do, your rank in your team, your organisation, or your industry. You will be able to apply what you learn, in any area of your life.

In most industries throughout the world, team members in all departments who are not in leadership positions, very rarely receive any leadership development. They are not taught these types of principles because they are "not leaders." Meaning, they don't have the leadership principle or title. However, I know that these teams are desperate for this type of development. They are desperate to make a difference in their team, their department, and their organisation. This is another reason I wrote this book, to give these teams what they crave…leadership development.

To build trust with a fellow colleague or teammate, you need to know two things:

1. Who they are (character)
2. What they know (competency)

Character development is far more important than competency development. However, most organisations have it the other way around. They focus on competency and not character, which is totally wrong.

To be a trustworthy person you need to know who you are (character). So, this book, and my other books will be more focussed on building your character so that you can learn to become trustworthy.

Go through each of these chapters thoroughly and apply what you learn. If you do that then you will know how to build trust with others a lot easier. However, being trustworthy is not the only principle you need to become a highly effective team player. There are a lot more, but being trustworthy is the first and most important. It is most important because it is the common thread through each principle in this book. It is also the common thread throughout yours and every other highly effective team in the world.

Other people will judge you on things you have done wrong. So, judge yourself on the things that you have done right. If you do that you will fall in love with yourself.

Have humility

Humility teaches us to put others first. Pride and ego manipulate us to put ourselves first.

When I moved to Scotland in 2011, I took a job as an engineering production manager. I was working shifts with Scotland's train operating company. My shift pattern was dayshift, backshift, and then nightshift. I was part of a 5-man team on a rotating shift. The engineering depot I worked at was based in Edinburgh, and it was an extremely tough place to work. Most of the engineers were good to work with, but the leadership team were not.

The depot manager was a very aggressive person and led his teams through fear. This style of leadership then filtered down onto his managers who reported into him. So naturally, they led their teams in the same way. At that time, I thought I was supposed to lead in that way too, so I did for a while. However, I had been studying leadership for 2 years at this point, and I knew in my heart I was leading the teams in the wrong way. But my head was telling me that if I didn't lead through fear, then I might lose my job. It was not a nice position to be in, and every day was becoming even more of a struggle.

After a while, I started to follow my heart and I tried to lead the teams through influence. I didn't want the teams to be in fear of me anymore. I tried to convince my teammates to do the same, but they were not willing to listen to me. After all, I was a 28-year-old inexperienced production manager from Liverpool. Why would they listen to me, or take my advice? They had such massive egos that it would have been difficult for anyone to change their ways. So, I just focussed on myself and the teams I was leading when I was on shift.

My relationships with the teams started to strengthen, my influence was increasing, and the teams were performing better. I even changed some of the processes within the leadership team, and my boss was really pleased with that. He even gave me some praise, which was rare. This was a great moment for me because I had changed my leadership style to one that I believed would work, and it was working. I believed in what I was doing so much and was driven to make it happen.

However, after about 2 years of me working in Edinburgh the performance of our trains started to deteriorate. We were getting a lot of customer complaints, and even made the evening news on more than one occasion. The blame for this was pointed to engineering. The directors accused our depot of not maintaining the trains correctly, hence causing them to break down in service. When this happened, this left a lot of passengers stranded at stations. It was not a good time for our depot.

The depot manager was "told" by the engineering director that someone had to be accountable. The engineering production management team were held accountable, and they then made me accountable. I was the newest member of the team, so it had to be me…right? Even though I was changing processes and leadership styles. My boss removed me from the team, and side stepped me into a different role within the same depot. He then replaced me with someone else, who didn't have as much experience as me.

But I didn't want to be part of that depot anymore. I was hurt by this move, and I wanted something different. So, I moved to Glasgow. I took a job as an engineering performance manager at the Glasgow railway depot. It was set up in the same way as the Edinburgh depot, but it was different.

The depot manager was named Stuart. He was a tall, well-built person and looked like a leader. Before I even spoke to him, I could tell he was the depot manager. He oozed confidence and charisma. When I spoke so some of the guys on the workshop floor, they all liked him. The engineering production managers liked him too. He wasn't fearful and didn't lead through fear.

When I had my first conversation with Stuart, he asked me why I moved over to Glasgow and took up this role. I explained what had happened in Edinburgh, and he already knew. Although he had heard a few different stories from different people. This was all gossip, and I didn't let it bother me.

Anyway, what I noticed most about Stuart in that initial conversation was how humble he was. He was the Glasgow depot manager and had nearly 200 people in his charge. However, he was a team player. He constantly talked about "our team", meaning the whole depot.

Following our conversation over the next few days I was in Stuart's company quite a lot. I attended meetings that he chaired and walked the workshop floor with him. He always stopped to chat with the teams, and not just about work either. Stuart is a huge football fan (soccer), as am I. So, he talked about football, family, and how work was going. He modelled humility effortlessly.

If we are not humble with our teammates, then we are not going to accomplish anything.

The depot manager in Edinburgh had too much pride to be humble. His ego was far too big, and that influenced the rest of his leadership team to be the same. This had a negative impact on me and the whole depot. If you have too much pride, you are concerned with *who is right.* If you have humility, you are concerned with *what is right.*

Fortunately, Stuart had a lot more humility than pride, and he influenced his teams to be the same. Stuart was one of the team. His influence was multiplied throughout the whole depot by his managers, team leaders, engineers, and the office teams. He focussed on doing the right thing, not on being right.

Stuart received the award of Leader of The Year throughout the whole of the Scottish Railways. So that was not just in engineering, it included all departments.

To be a highly effective team player, you must be focussed on "***WE***" not "***ME.***" You need to put the team before yourself. That is what humility is all about. Team players who have too much pride or ego are focussed on "***ME***" not "***WE.***" These team players are usually low performing and put themselves before anyone else.

A low performing team player with lots of pride and ego are oblivious to the damage and suffering they are causing within their team. They are so focussed on themselves that they don't even care about their teammates. When a team resolves around a low performing team player, they don't accomplish a lot. They are held back by the flaws in the low performing team player's character.

A low performing team player will look at someone who has humility as weak. Their ego makes them think that they are the strongest member of the team. When in fact, it is the total opposite. They don't realise that their influence decreases more and more as they continue behave in this way. They don't know any other way, and therefore don't know that they need to change.

A person with a lot of pride and ego are exposing themselves to the rest of us as being insecure. An insecure person does not have humility. They know within themselves that they are insecure, so they try to mask this insecurity by having an ego. The ego is not their true self, it is an act.

Highly effective team players do not need to mask who they really are. They are their true selves every day. They know who they are, and that is why they are highly effective. They take pride in helping others and giving the team credit for any achievements. They put the team before themselves. They will raise the team up before raising themselves.

Stuart is a highly effective leader, but he is also a highly effective team player. He put the team before himself, he helped them as much as he could. He gave credit where it was due and never took any credit for himself. He spoke highly of his leadership team, his engineers, and his office teams. I am proud to call Stuart a friend.

If you are a humble person, then it shows that you have a strong character. We must constantly be working on and building our character. To have humility, you do not need to be in a leadership position or have a title. To have humility, you must put others first.

We must work hard every single day to be our true selves. This is not easy to do. I would rather be myself than have an ego and try to be somebody I'm not. I would rather be open with people and for them to be open with me. What would you rather do?

When you take responsibility, you are on the path to becoming highly effective

CHAPTER 2
POSITIVITY

Take responsibility

When we are successful in life, we must remain humble. When we fail in life, we must take responsibility and accept it. These are very difficult traits, but they are essential.

In the previous chapter we discussed humility, and why we need to have humility to be a highly effective team player. We need to take responsibility and *make the decision* to have humility. That's what humility is, a decision. There are certain circumstances that make the decision easier or harder. For example, if you haven't done a lot in life and you have no ambition to achieve a lot in the future, then it is a lot easier to have humility.

However, if you have achieved a lot and you have ambitions to achieve a lot more, it is much harder to have humility. The reason for that is, success can go to our heads and people can become cocky or arrogant. They let it go to their heads basically. So, we need to be strong in these situations, and making the decision to have humility will give us strength.

In my experience of working in the engineering industry, I have found that a lot of people do not let success go to their heads. They make the decision to have humility. I believe this is true in most industries, especially with the teams who are not in leadership positions. It seems to be when people become supervisors, team leaders, or managers that they let things go to their head. I have worked with some very arrogant and egotistical leaders before, and they were not followed by anyone. I have also worked with leaders who had humility, and they were followed by everyone.

A highly effective team player is one who will always have humility, no matter how successful he/she is. They go the extra mile when required, and they help others to do the same, and they show them the way to humility. I have always worked very hard to not let any successes I had go to my head. I don't like arrogance, so I would never allow myself to get like that and I would always have humility. I find that it is my responsibility to make that decision.

Being part of a team and being a highly effective team player, it is our responsibility to determine just how effective we will be. What we choose to value is also our responsibility, and those values will either keep us on the right path or they won't.

The decision to take responsibility is not an easy thing to do. It doesn't matter if you are in a leadership position, or you are one of the team. However, taking responsibility is a lot more difficult for the low performing team players, just as the decision to have humility is difficult for the high performing team players. It doesn't matter how we look at it, this is hard work. But we must not give up on making these decisions.

If we decide to not take responsibility, then this will create a lack of trust within our team. This is common when it comes to low performing teams and insecure team players. When it comes to taking responsibility, they duck and dive anywhere they can. When it comes to being allocated a task, they will always have an excuse; it is not their job. They don't get paid enough. They don't know how etc.

A highly effective team player would do the opposite. If a job needs doing, they will volunteer first to do that job to the best of their abilities. Especially if a low performing team player decided that they can't do it. This is an opportunity for the highly effective team player to raise their status within the team and increase their influence. So, the secret is, when a low performing team player doesn't want to step up, jump on the opportunity and step up instead of them.

When a highly effective team player does step up, they immediately increase their influence and trust with the team. They have made the decision to take the responsibility to do the task. Now it is their responsibility to complete the task to the best of their ability and achieve the desired results. If they can achieve these results, then their influence and trust will increase even further. If they cannot, then it may cause a lack of trust with the team, and with their leader.

However, it is rare that a highly effective team player will not achieve at least the desired results. They know that to keep increasing their influence and trust with their leader, they need to help the leader by focussing on the task. They will not stop focussing until the desired results are achieved. The leader of any team has total responsibility for every result, so they focus on helping the team achieve them. If they don't then they risk being replaced by another leader. A highly effective leader does not want to see that happen.

As a highly effective team player's influence and trust increases with their leader, and they continue to achieve results, they will start to stretch their influence. They will begin to influence leaders in different teams and departments within the organisation. When they decided to take responsibility and have achieved great results, they start to build a reputation for themselves. As their reputation develops, they start to be relied upon a lot more to give their opinion on certain things.

At this stage, the highly effective team player is still one of the team, but they are performing at a much higher level. They are especially at a much higher level than the low performing team player. With the highly effective team player's newly found reputation, they start to influence the direction that the team are headed. They also start to influence the decisions that the leader makes.

If you want to strive towards becoming a highly effective team player, then you must start behaving and thinking like a highly effective team player. For example, a highly effective team player will wake up in the morning and be hungry to seek out the opportunity to increase their influence with their leader, and their team. A highly effective team player is driven to achieve great results and will not stop until those results are achieved. A highly effective is not short sighted, they see the vision for the team just as good as the leader does. Most of all, a highly effective team player will always take responsibility.

Think about the team you are in now. Do you have team players who are willing to take responsibility? Or do you have team players who are not, and will make excuses? What kind of team player are you right now?

If you were part of a team of low performing team players, where everybody was insecure and not willing to take responsibility for anything. How would you feel? Frustrated? If the answer is yes, then this would be your time to take responsibility, increase your influence with your leader, and achieve great results. If you were to do this then the leader's frustration would ease, and the team would see you as a role model. Taking responsibility would send a message to the other team members to follow in your footsteps.

If you were part of a team of high performing team players, where everybody was secure and were all willing to take responsibility for everything. How would you feel? Great. If the answer is yes, then every task would be completed with great results. There wouldn't be any frustration in anybody, just happiness. Everybody would find it easy to focus on achieving great things. Everybody's influence would increase with the leader and each team member. However, this would be yours and the team's time to influence the other leaders, the other teams, and the other departments. If you were to do this, your influence will multiply, your reputation would continue to grow, and so would your followers. It is a win-win situation for the whole organisation. Would you like to be part of this team?

When it comes down to it, the only difference between the two teams we have just discussed is; one team is negative, and one team is positive. The low performing team players are negative and have a bad attitude. The high performing team players are positive and have a great attitude. As it is a decision to take responsibility, it is also a decision to be negative or positive. If you want to be a highly effective team player, make the decision to be positive and have a great attitude.

It is never wrong, to do the right thing. We must take the responsibility to do the right thing and take ownership for our actions.

Master your emotions

Do you want to master your emotions, or do you want to be mastered by them? It is your choice.

Have you ever noticed or investigated why people get fired from their jobs? I have both researched this and witnessed it on 3 different occasions. The main reason people are fired from their job is not because of their competency (what they know), it is because of their character (who they are). So, basically, they are fired if they either behave negatively, have a bad attitude, can't control their temper, or aren't trustworthy.

However, when these people are hired, they are hired because of their competency (what they know). Therefore, it would make sense to help our teams build on their characters at work…wouldn't it? That way, they have less chance of being fired. It would also show our teams that we do care, and that people's emotions are important to us.

Emotions are important because within a team, one person's emotion can infect the whole team. Especially if that is a negative emotion. When I was an engineer and part of a team, negativity was very common. A negative attitude only had to come from one team member, and suddenly everybody had a negative attitude. It seemed a lot more difficult to infect the team with a positive attitude.

You can witness this for yourself.

When you are next in work and you are in the breakout area or the canteen, try to listen in on a group's conversation. Listen out for what they're saying and gauge what the general mood of the group is like. Is it positive or negative? It will most likely be negative because negativity is a lot more contagious than positivity. Do the same thing when you are part of a meeting. What is the atmosphere like in the meeting? Is it positive or negative? Make a note of it and see if there is a common theme between the different meetings you go to.

When you are part of a group discussion, or in a meeting that is mostly negative, do your best not get dragged into the negativity. This is not easy to do, but it is imperative for you to remain on the right path to becoming a highly effective team player. If you become a negative person in your workplace then you will have a huge impact, but the wrong impact. You will infect others with your negativity. You will hinder your chances of new opportunities. You may not receive a pay rise. You most certainly will not be promoted. But, most importantly your influence will decrease.

When a highly effective team player is part of a negative group discussion, they either remain silent, or they try to spread their positivity through the group. So, try to do the same and either bring positivity into the conversation, or just remain silent. Observe everyone in the group while the negative discussions are taking place. Spot who the ring leaders are. It is highly likely that they are low performing team players, as this is a very common trait of theirs.

Over the years of working in the engineering industry, I always found that the most negative people were the ones with the loudest voices. Do you find that to be the same in your industry too? They shout the loudest because they want to bring everybody around them to their level. They enjoy being the ring leaders of negativity. They want everyone to be just as low performing as they are.

I've mostly tried to spread a positive message among teams throughout my career, whether I was the leader or not. A lot of the time, my positive message was ignored, or replaced by a negative message. Usually, the negative message would come from the low performing team players.

Highly effective team players are sadly among the minority in the workplace.

So, don't get discouraged if you are outnumbered by low performing team players. Keep spreading your message of positivity. It is times like this that you will truly be building up your character. Emotions such as anger and frustration cannot come out at the same time as a positive emotion. When you are positive you are pushing anger and frustration away. When you are frustrated and angry, you are pushing positivity away. Overcome anger and frustration and remain positive.

In the early days of my career when I was in my 20's and started studying leadership, I would do things that would keep me positive. One of those things was to go onto social media and join leadership groups. I wouldn't join so I can chat every day. I would join with the idea that I could read the positive stories from leaders. One day while reading these stories, I came across Zig Ziglar. He posted a picture with a quote on it that read *"Do something today that your future self will thank you for."* I loved that quote.

So, what I did was print this quote off and stick it up on the notice board in the office. I was working for Liverpool's train operating company at the time.

It remained up there for as long as I worked there, which was 6 years. I was searching through these groups daily for more positive quotes. I remember one from Simon Sinek which said, *"true strength is the courage to admit our weaknesses."* Again, I loved this quote. But, instead of putting it on the notice board, I emailed it to everyone in our engineering team. Some people laughed, but most replied to me saying thanks. So, I continued to email our team similar quotes from the groups I had joined.

Among the team you are in at this very moment, what is the general mood of the team? Are they constantly negative or positive? What is your general mood when you are at work? Are you constantly negative or constantly positive?

I have worked with many negative and low performing team players, but none more so than when I worked in Edinburgh as engineering production manager in 2011. There was one person whose name was Tam (His real name is Tom, but in Scotland they say Tam). He was a very hard worker, but he was often frustrated and angry. His attitude was very negative towards his team leader, towards me, and towards his teammates.

I sat him down to discuss this with him. He told me he was frustrated and angry because he felt nobody would listen to him. He told me about some of his ideas, but he felt he couldn't share them because nobody would take any interest.

I asked Tam how he knew that nobody would listen to his ideas. He replied, *"I don't know it for sure, but I just know."* He felt that his team wouldn't listen, so he was just assuming. I told him to give it a try with his team leader, to share one idea with him. Then, I printed off the Zig Ziglar quote *"Do something today that your future self will thank you for."* He laughed as I expected, but he took it from me.

When our conversation finished and Tam went back to work, I know he was thinking differently. He was thinking of *why* he hadn't shared any of his ideas. The truth is, he wasn't angry and frustrated at his team. He was angry and frustrated at himself. But his negative attitude was pointing the blame towards his team.

The next day, Tam approached everyone with a smile and a hello instead of a frown and a growl. He was deliberately being positive all day. He found it hard, but he did it.

The next shift that Tam was working, he asked his team leader if he could have a word. He shared the idea of job swapping within the team. He thought that by doing this, it would develop the rest of the team to become better engineers as they wouldn't be doing the same thing every day. It also had the potential to improve teamwork, help the team to build better relationships and trust, and ultimately make the team happier.

His team leader then came to me with Tam's idea, and I said, "*Go for it!*" So that's what they did, and it worked.

Tam's idea helped build the morale of the team, and they showed their gratitude by taking Tam out for a beer. Tam's attitude totally changed. He was smiling a lot more, and he was no longer angry or frustrated. His positivity shone through, and he became a highly effective team player.

If your team's morale is low and nobody is doing anything about it, take the responsibility to improve it yourself.

Embrace change

Live for the future, not your past.

Organisations talk about change a lot. It is often described as continuous improvement, and some will have a continuous improvement department. I went for a job in 2013 for a continuous improvement manager but didn't get it. Some people love when things change, but a lot of people hate it. They run so far away from it and hope it doesn't catch them. The thing is it always does catch them.

A highly effective team player embraces change. They see it as an opportunity to grow and *continuously improve.* A low performing team player will reluctantly go through change. They do not embrace it or see it as an opportunity.

Whenever a new change is discussed, you will always see resistance from the insecure, low performing team players. Then, when the change happens, the same low performing team players will do their best to block it. I have experienced this many times both as an engineer and an engineering manager.

However, I have also experienced that highly effective team players will do the opposite. They understand that for change to happen, you need to be flexible. You need to have a positive attitude when it comes to change. They don't seek out the negatives to stop the change from happening. They understand that if you remain positive then you can make the change happen. They also understand that you can make the change benefit both you and the team, and not do damage.

I really hope you are beginning to realise that the principles I am sharing are all interlinked. Every single one of them can be combined when they need to be. They are designed for you to become a highly effective team player, so you can have the most positive influence possible on your team. These principles will also help you to multiply your leader's influence, by helping your team increase their influence too. Keep this in mind throughout the rest of the book.

It is the role of a highly effective leader to lead their team through change. But it is also the role of a highly effective team player to do the same. They can take the lead from the leader, and multiply the leader's influence, by increasing their own. When they do that, they will also be helping the team to do the same. As the highly effective team player is increasing their influence, they are building momentum. As the momentum builds, the progress of the change also builds. Change is a very powerful way for a highly effective leader, and a highly effective team player to work side by side. It is an opportunity for them to raise the team to a new level and is one that should be embraced.

You are on the path to becoming a highly effective team player. What are you going to do when change comes around in your organisation? How are you going to embrace it? How are you going to support your leader? How are you going to help your leader to lead the change?

Keep thinking of change as an opportunity to grow. Look at change as a chance to become a more valuable team player. The most valuable team players are the ones who lead their teams, and their organisations through all types of change. They do this every time a change comes around. How valuable do you want to become?

In the team you are currently a part of, are you a team player? Do you step up and play for the team, or do you sit on the side-lines? I can tell you; it is very boring sitting on the side-lines. You will have a much better time as one of the players on the team sheet. How about being the first name on the team sheet? That would be exciting. Highly effective team players never sit on the side-lines, they are always right in the middle of the action. They lead the way when it comes to change and will do everything, they possibly can make change happen.

Earlier in this chapter we discussed Tam from my days in Scotland, and how he changed his attitude from negative to positive. What he realised was that to be negative or to be positive is a decision we make every day. Not many people know that, but it's true. We decide *"Today, I'm going to have a negative attitude"* or, *"Today, I'm going to have a positive attitude."* What decision did you make today?

When a person decides to change their attitude from being negative to being positive, things around them will start to change too. They will start to look at things differently and will attract positivity into their lives. That doesn't mean there won't be challenges, issues, or problems because there will…every day. But a negative attitude will not be one of those problems anymore.

The more we can hold our positive attitude, we are breeding a behaviour that others will want to emulate. So, they will become a lot more positive too. Especially, if we are part of a team that is full of negative attitudes.

I do my best every day to keep a positive attitude and mindset. Believe me, I find it difficult sometimes, as do a lot of other people. What I do to try to maintain positivity is; I look within myself a lot. If I feel the urge to blame someone for something, I resist it by looking within. I see this urge as an opportunity to work on myself and discover how I can stop these urges of blame. I do the same for any other negative urges I am feeling.

A lot of people all over the world tend to look outwards rather than inwards. But, if we want to change and be positive, then we need to change too. We need to start looking inwards a lot more than we look outwards.

If you are reading this, and you are a person who doesn't really like change, then don't worry. I am going to help you to think differently.

I was the engineering technical manager at a train operating company that ran through East Anglia in the UK. During that time, I became part of a team who were bringing in a big change. The team consisted of depot managers, planning managers, and production managers. We were all the same level. The change was to bring in a new model called MPRM. This stood for: Maintenance, Planning, and Resource Model.

The purpose of this change was to improve how the engineering department did business basically. It would need to involve every single person from within the department, which consisted of hundreds of people based all over East Anglia. So, this was no small change. In fact, it was huge.

The first person I approached about this change was one of my technical engineers called Stuart. He was based in Norwich. We greeted each other with a smile, handshake, and hug, as I did with almost everybody. I said "*Stuart, I am part of a team that are bringing in a huge change that is going to improve our processes, maintenance, resources, planning……everything. I am hoping you can be one of the change leaders for our team.*"

I could see in Stuart's face that his cheeks were becoming a lot redder. Stuart was a brilliant engineer, but he was set in his ways, and didn't really like change. Especially any changes that were related to his job, as he prided himself on what he did. But I explained *"I need your help, you're the best to lead this change for Norwich. I need someone with a positive attitude."* To my surprise, Stuart replied *"I would love to."* I said, *"Thank you so much, I'll get back to you."*

Stuart had a reputation of being very stubborn when it came to change, but he made the decision to be positive that day. That made me very happy. He then went on to lead the change and is even leading the change to commission the new trains for his depot. All because he decided to change his attitude.

That's what change is…a decision.

Change is not easy, it is difficult. But is it worth it? Most certainly. If you find it difficult to embrace change, then make the decision to change, and then embrace it.

Try changing, you will not regret it.

How do you see reality? Do you face up to it, or do you avoid it? The consequences of avoiding reality are much worse than facing it.

If you want to improve your results,
improve your focus

CHAPTER 3
FOCUS

Be laser focussed

Winners focus on winning. Losers focus on trying not to lose. Are you a winner or a loser?

Being advised that you need to be focussed, or be laser focussed is one thing. But what are you being advised to be laser focussed on? I am a huge football (soccer) fan, and my favourite team is Liverpool. Throughout my life, Liverpool have always been focussed on winning. They want to win every match and win trophies at the end of the season. There are other teams who play in the same league as Liverpool, who focus on trying not to lose. I won't name any names here (Everton).

The difference is in terms of football; when you are laser focussed on winning, you know exactly what the vision is. You also know how to get there. When you are laser focussed on trying not to lose, you are thinking of either winning, or drawing. So, the vision is very hazy and unclear.

If we think in terms of our working world; being laser focussed on winning, is the same as being laser focussed on what the vision for the team, and the organisation is. We are clear on where we want the team to get to, and how we are going to get there. Being laser focussed on not losing, we are not quite sure on what the vision for the team, or the organisation is. We also don't know how we're going to get there.

Highly effective team players are always laser focussed on the vision for the team. They think in terms of the big picture. When they start work in the morning, they have a purpose, and they want to make a difference every day. In football terms, highly effective team players are laser focussed on winning, and they will do anything they can to help their team win. They go the extra mile and go further than what is expected of them. How much are you willing to be laser focussed on winning?

Low performing team players are not laser focussed on anything usually. They don't know the vision for the team, and don't care about the vision for the team. When they start work in the morning, their objective is to be assigned as little tasks as possible. They don't want to be part of the team and will not help the team win. They don't have a purpose and are focussed on trying not to lose. Low performing team players look at what they do as just a job and will do the exact minimum required to get the job done. How much are you not willing to be laser focussed on trying not to lose?

Over the years I have led some great teams, and some not so great teams. The great teams had a vision to strive towards, and most members of those teams knew what it was. So, they were laser focussed on striving towards achieving that vision every day. There were one or two team members who were not, as they were mainly laser focussed on themselves.

Members of the team who only focus on themselves, are low performing team players. They are usually part of a team because they were told to be, not because they wanted to be. The highly effective team players are there because they volunteered and wanted to make a difference.

If a person is part of a team because they are told to be, and they don't want to be there, then that will drag the rest of the team down. Their focus won't be on the team's vision, but mainly just to get through the day with doing as little as possible.

It's possible that you have experienced this in your career. You may have been part of a team with people who were told to be there, and people who volunteered. Have you ever been part of a team that you didn't want to be a part of? Have you ever volunteered to be part of a team? If the answer is yes to both questions, then you know the difference in feelings. One team you were excited and driven to be a part of. One team you felt miserable and wanted to be somewhere else. I know the feeling because I have experienced the same in my career.

If you were in the position to lead a team, which type of team members would you want on your team? Would you want the team members who were told to be there (low performing team players?) Or would you want the team members who volunteered to be there (highly effective team players?) The answer is easy, highly effective team player, right?

When volunteering to be part of a team, or part of a group, you are giving yourself a great opportunity. You are willing to accept the responsibility to make a difference, and you are giving yourself the opportunity to raise yourself up. People who volunteer to be part of teams are usually highly effective team players. So, they will know, and be laser focussed on the team's vision. They will also know how to strive towards achieving that vision.

I would like you to look within yourself for a moment and be 100% honest with yourself. If a new project came up in your workplace, and had the potential to be a huge benefit, would you volunteer to be part of the team? Or would you wait to be told to be part of the team? Whatever your answer may be, will your influence increase with the highly effective team players you work with? Or would it decrease? What do you think the difference would be from you volunteering to be part of the team, and being told to be part of the team?

If you are thinking that you would like to be on the project, but not part of a team, then I urge you think a bit deeper. It is a huge project and will benefit the organisation greatly. This cannot be carried out by individuals; it must be a team. If you are not willing to be part of a team, then you are a low performing team player, and focussing on not losing. You will soon be bypassed by the volunteers who are highly effective team players, and are laser focussed on the vision, and winning.

You are part of the organisation you work with, that is one team. You are part of the department you work with, that is one team. If you are part of a project, that is one team. If you work alongside a different department, that is one team.

You are part of many teams that form one organisational team. So, you cannot dismiss being part of a team. All successful people became successful because they were either part of a team, a purpose, or a cause. They had help from a team. To be truly successful, you cannot do it all on your own.

I didn't realise the team scenarios above until I became a project manager, when I worked in Liverpool for their train operating company. The reason I didn't realise these scenarios was because for years we were not all focussed on the vision. We were focussed on trying not to lose, or just getting through the day with no issues. It wasn't until we had a new engineering director, Kevin, who turned us around and we focussed on his vision for engineering.

Whatever organisation you work with, you are part of a team of teams. If you don't focus on the vision for the team, and the organisation, then you will be a low performing team player. If you do focus on the vision for the team, and the organisation, then you will be on the right path to becoming a highly effective team player. Make the decision of what you want to be *laser focussed* on. The team, or yourself?

If you focus on yourself, then you are letting the team down. If you focus on the team, then you are helping to raise them up.

Highly effective team players are laser focussed on winning. They want to win. They know how to win. They will do everything they can to win and won't stop until they do. Is this you?

What you choose to ignore in life, could have serious consequences in your future. Be laser focussed on the right things, for the right reasons.

Have discipline

What does your today look like? Make better decisions today, so your tomorrow is better.

Highly effective team players have great discipline. When they say what they're going to do...they do it. When they say when they're going to do it...they do it. When they say how they're going to do it...they do it. They do not sugar coat anything or tell people what they want to hear. They are straight and true with what they say, and they are 100% truthful in what they say.

I have been part of lots of teams, and a lot of the people I have worked with have talked the talk, but they didn't walk the walk. Meaning, they said that they were going to do something, and it was very rare that they would follow through. They have even committed to something, and still didn't follow through. You have probably worked with, or still work with people like that. One of these people may even be your boss, or your boss's boss. The point is, if you make a commitment, it doesn't matter what level you are at. Whether you're the CEO, or the cleaner, it has nothing to do with your position or title. It has everything to do with your character.

If you have worked with a person who broke a commitment they made to you, did you continue to trust them?

Thinking about yourself, and your commitment to keeping commitments, ask yourself, "Am I renowned for breaking commitments?" Being totally honest with yourself, "How often do I break commitments?" "What has the impact been on my team, colleagues, friends, and family?"

Highly effective team players will always ask themselves, "Is this commitment I am about to make going to build trust with the team, or make me untrustworthy?" A highly effective leader is focussed on the team, and the impact their commitments will have on them. When you are making a commitment, you are revealing to the team, or your friends and family what you intend to do. If you keep that commitment, or if you don't, then you are revealing to the team, your friends and family your character level. Character is the difference between a highly effective team player and a low performing team player.

If you make a commitment with a person, and you keep it, then you are building trust with that person. If you make a commitment to a person, and you break it, you are becoming untrustworthy. Making commitments to people is very serious, so ensure that you treat your commitments in the same way. What you commit to is very important to the recipient of that commitment. Put yourself in their shoes and think how they would feel if you were to make a commitment and break it.

When making a commitment, you ***MUST*** have the discipline to follow through. If you don't then you are not only cheating the person you made the commitment to, but you are also cheating yourself. You will start to lose trust with yourself and fall less out of love with yourself. You are not only committing to another person, but you are also committing to yourself. So, ensure that you follow through.

Not having the discipline to follow through with a commitment is one thing, and you will feel bad initially. But the feeling of regret afterwards is far, far worse. I know this because I have broken many commitments throughout my life that I truly regretted. You may have too, so you may know the feeling I am talking about.

When I was in my teens, and I broke commitments with my friends, or members of my family, I didn't really feel anything. I didn't know that I needed discipline to follow through. I didn't even know what I was committing to was even a commitment. However, as I grew older, the commitments I was making were much bigger. I knew that they were commitments, and that I needed to be disciplined to follow through with them.

I remember breaking a commitment with a team member in one of my first management roles, his name was Davey. I was supposed to do his appraisal, and I told him I was going to help him put an action plan together for his development. But I forgot. So, another manager did his appraisal, and did nothing about his development plan. My boss asked me why I didn't do it, when I had committed to them both that I would. I apologised and told them I just forgot. A lousy excuse on my part.

My boss took me aside to have a word with me. He said, "Tom, I don't expect this from you. Put yourself in Davey's shoes. How would you feel if I was to make a commitment to you, then didn't follow through, and just told you I forgot?" "Do you not feel any regret for what you have done?" "Davey is very upset. You broke your promise to him, and you need to make it up to him." His words really did hit home with me. I was not usually one to behave like this, but I did.

The regret I felt was very heavy on my shoulders, and I needed to rectify this. So, I went to Davey and had a very in-depth conversation. I apologised to him and told him how much I regretted missing his appraisal. I was sorry for how flippant I was with him in the aftermath. I made it up to him by re-arranging our appraisal, and I helped him put together his development plan. I agreed to mentor him on his progress and help him anyway I could. The plan was about 6 months long, and he learned new engineering skills. He even increased his workload and pay with his new skills. I was very proud of him.

It is important that if you do break any commitments, then you make up for them if you can. You don't want the feeling of regret on your shoulders because it is very heavy.

If you are now thinking of your own regrets in life, it is a good thing because you are acknowledging that you have broken commitments. However, my advice is not to dwell on them. If there was something that you committed to a long time ago, but didn't have the discipline to follow through, then there is always today. Use your past to inspire you today. Work on yourself every day to become the best person you can be. Make that commitment to yourself.

When making this commitment to yourself, remain positive with it, and follow through. To be positive every day is difficult, and it takes discipline. As with a lot of the principles in this book, discipline is a decision. We must decide each day to be disciplined or not. When making that decision, take small steps. Don't take huge strides. Focus on the small things first before you move up to the big things.

The more you focus on the small steps, you will eventually make the big strides. Be disciplined to just focus on the small steps, as this will help you be positive too. Stay the course, and you will soon develop more and more self-discipline to take bigger steps.

As you develop this self-discipline and you remain positive, people will start to take notice. When you are at work, you will receive compliments on how you are brightening people's days. You will influence your team a lot more, and they will start to follow your lead.

You will also get negativity from some people, and they may avoid you. But don't let that stop you. You are on a mission to be self-disciplined, be positive, and become a highly effective team player. Your leader will support you and will know that you are doing the right thing for yourself, and the team. Let the low performing team players be negative. Then just watch how they change when everybody else is following your example. Remember, *behaviour breeds behaviour.*

As you go through these principles, you need to be disciplined with them all when you start to apply them. I believe you can.

What culture does your organisation operate? Whatever it is, you can be the catalyst to improve it to a culture of discipline. If you can do that, the people will see their jobs as a responsibility.

Take action

It is always good to plan action. But it is far, far better to take action.

When we discuss being deliberate and being disciplined, these are very similar traits. However, when we break them down, it can be determined that they are different traits. When we are being disciplined, we are committing to taking action on something that will benefit ourselves and will benefit others. When we are being deliberate, we are committing to taking action so that we can achieve a specific goal, or objective.

Part of everybody's role when I worked for the train operating company in Liverpool, was to help improve the performance of the trains when they were in service. That was the philosophy throughout the company, not just for the engineering department where I worked. In 2008, a new asset management strategy was to be implemented by the engineering director and his team. I wanted to be part of his team, so I was one of the first engineers to volunteer. This was an excellent opportunity for me to be very deliberate in my personal growth, so this was my focus. If I was to work on myself, and help the team, then I had to make a huge commitment that would require a great amount of discipline. You will learn as you continue to read, the traits that are required when it comes to making this commitment and keeping disciplined. The team and I were taking on a huge challenge, probably the biggest challenge any of us had faced in our careers so far.

As I was one of the first to volunteer, I was chosen to lead some of the activities that would form part of the asset management strategy. The duration of the strategy was a two-year process, so it was a huge commitment from everybody. I made the commitment to my engineering director and to the other leaders that I would take action every day. This required me to be deliberate with the specific actions I took, every single day. My leadership skills required improvement, so I was reading leadership books, and other materials every day. I still do to this day. I had little experience in asset management, so I read books on asset management strategies. I had to keep disciplined when it came to my reading, and I made sure I wouldn't procrastinate or cut corners in any way. I made the commitment, so I needed to follow through.

If you are part of a team and would like to learn how to lead your team, even though you are not the leader, then read my book; *Influence is Leadership.* This book focusses on helping you and your team to reach your full potential.

When I started leading some of the activities as part of the asset management strategy, the teams knew that I didn't have a lot of leadership experience. So, they played on that quite a lot to begin with. However, it was my job to change how they thought about my abilities. For me to do that, I had to build trust and build relationships with the team. If I was to build trust, then I would have to be deliberate and know the process inside out. I had to know the activities and what was required better than anyone else. So that's what I did. I made the sacrifices to learn as much as I could and become very knowledgeable. I had to know and understand the requirements of the team, so we could work together to achieve the desired results. As a team, that is exactly what we did.

From 2008 to 2009 the asset management team achieved an improvement in train performance of 40%. We were very proud. We also seen an improvement in our own engineering department's performance. In the end we achieved great things as a team, and we were recognised for it.

As with most teams, we didn't just have highly effective team players. We had low performing team players too. Can you guess what they focussed most of their energy on? Let's just say that if we had followed the advice of our low performing team players when we started out, then we would not have achieved anything. In fact, the performance of the whole department, including the trains would have decreased. They would constantly be telling me that the asset management strategy was a bad idea. That I didn't have enough experience to help lead the activities. They wondered, *"Why we needed to change things when we were already "doing ok?"* Well, I didn't want to just do ok, and neither did the rest of the team. We wanted to achieve amazing things and have an amazing time while doing it.

Thankfully, nobody listened to the low performing team players, especially the highly effective team players. Whatever the low performing team players were trying to sell, nobody was buying from them. The team instead decided to take responsibility for what we were trying to achieve, and we all became deliberate in ***TAKING ACTION*** to make things happen. The improvement of 40% in performance of the trains resonated throughout the whole business. Our engineering department was growing in performance, and we were growing because of our achievements.

Unfortunately, when a new strategy is being introduced in any business, there always seems to be an *"us and them"* mentality within the teams. Throughout my career, I have seen this so often. The *"us"* side of the team are the highly effective team players, and the *"them"* side of the team are the low performing team players. The high performing team players are the people who support the new strategy. They want to share the vision, and the mission that the leader has set. The low performing team players are the people who don't support the new strategy. They want to do the exact opposite to the leader's vision and mission.

I have yet to be part of an organisation where everybody within the team are all pulling in the right direction. Maybe you can be the first to make this happen in your organisation.

If there are any process improvements, or new strategies about to be implemented within your organisation, then make sure you are part of it. Volunteer and make the same commitment to lead part of the activities for your team. When you do that, you are proving to the leaders that you have a positive attitude about the new process improvement. You will also be taking responsibility and being deliberate by taking action.

When I decided to be part of the asset management strategy I discussed earlier, I could feel my mindset changing. I was a lot more positive and could feel myself becoming a highly effective team player, and leader. I found it easier to focus on what I and the team were striving to achieve. Before I volunteered, I felt like I was doing the same thing each day and going through the motions. I was not being deliberate, every day. But now, I had a new purpose, and I became a lot more deliberate.

The team and I were becoming a lot more productive too, and we started doing things that would save time in our areas of work. When somebody had a new idea, we would embrace it and try it out. If it didn't work then we would move on, if it did then we would continue with it. There was no more discarding of ideas before we even tried them out. This made the team and me a lot happier. When we had our team meetings, they became very efficient because everyone knew what everyone else was doing. We were working as highly effective team players, and the low performing team players were missing out.

It's amazing to think what it would be like if everybody in your organisation, and everybody in the organisation's I've worked in previously had the same kind of attitude and was deliberate. Who knows what we all could achieve? What if our countries leaders had the same attitudes and ways of thinking? Our world would be in a much better place. Being deliberate is an easy concept for us to take action on things. It is up to you to lead your team, and your organisation through this.

So, ***BE DELIBERATE*** and ***TAKE ACTION***.

You may not have all the answers, but don't let that stop you. Make the most of what you have, and don't give up.

Do you have a sense of purpose? A sense of purpose allows you to focus

CHAPTER 4
PURPOSE

How driven are you?

Good things don't come to those who wait. Good things come to those who are driven enough.

Everyone throughout the world is driven by different things. For example, I am driven by helping others. Hence, I am writing this book and my other materials to help people with their leadership, and with their personal growth. When people are inspired, they are driven internally, from how they are feeling inside of themselves. When people are motivated, they are driven externally, so it requires certain actions from others that will motivate them. Highly effective team players don't have any issue with being inspired and driving themselves from within. Low performing team players are not driven from within, so they need the actions of others to motivate them.

Highly effective team players in every industry in the world are driven in the same way. The principles you are learning as you go through this book, they are proven and tested to work. Highly effective team players are aligned with these principles and embrace them all the same. How we can differentiate between each highly effective team player is the results they see both individually and as a team. The description of how/where they applied these principles will be different with each highly effective team player. The principles are the same, but the application is what is different. What inspires and drives each highly effective team player will also be different, as this is very personal to the individual.

Do you want to become a highly effective team player? How driven are you right now? What *being driven* means to a highly effective team player is; *don't wait for things*. Don't wait for someone to tell you what to do. Don't wait for things to happen within your team or organisation. Don't be afraid to say what you think, at the right time.

When you become a highly effective team player, you will be driven to be proactive. So, if you think something should be done, then do everything you can to get it done. If you have an idea that you think will work, talk to your boss and recommend the idea to him/her. Don't give up until the idea is at least tried out. If it doesn't work, move on. If it does work, then do everything you can to get the idea implemented. The most important thing is ***DO NOT GIVE UP!***

A low performing team player will do everything they can to get away with as much as possible. All they want to do is get through their hours for the day, and then go home. They will not think for themselves or be proactive. They will wait to be told what to do by their boss, or someone else's boss. I have heard people say in the previous places I have worked, *"I'm doing nothing today, Tom you are doing all the work today."* So, I asked *"Why is that? Why am I doing all the work, and you nothing?"* They replied, *"Because I don't want to be here. If I don't have to do anything, then I won't."* What an attitude to have. I was only young then. But, if I had a person like that on my team when I was a leader, they wouldn't be on my team for much longer. They were not driven whatsoever. They didn't care. They had no purpose. It is very sad to think that there are people like that in our workplaces.

As a highly effective team player, you will have the drive to improve things within your team and your organisation. This is the type of person I would want on my team. I want the type of person who thinks ahead. So, if they are currently working on a task to improve things, they are already thinking about the next improvement, and then the next improvement. They will have a vision of what their improvement ideas look like and be driven to get their ideas implemented.

Would you want a highly effective team player like that on your team? Or could you become that highly effective team player? Do you have the drive for it?

When I was part of the asset management strategy team working for Liverpool's train operating company, I led several activities. 50% of the activities were done as part of a team that I led, and 50% of the activities I did as an individual that didn't require others all the time. However, when working on my own, I didn't see it as actually working on my own. I always felt part of the team, and I wanted to make the best of the new asset management strategy. I had a purpose, and I wanted to make as big an impact as possible.

I felt like a highly effective team player, and eventually I did become a highly effective team player. This wasn't trained into me, or a motivation from somebody else. This was inspiration from within. I was part of something that I knew would make improvements on the performance of our trains, and most importantly our people within the organisation. I was driven to ensure that these improvements happened, and that everyone within our engineering organisation benefited from it.

I made sure that the other members of the asset management strategy team had the same opportunities to put their mark on it. I wanted them to make a difference too. So, we swapped roles. Some days I would be leader, and other days someone else would-be leader, and then someone else. It was a great way to share ideas, experiences, and it became a lot of fun. We were all driven to make this strategy work, and it did.

We improved the performance of the trains in service to 40% from 2008 to 2009. This was one of the biggest improvements that the company ever made, especially in the engineering department. Everybody was happy at the time, and things were going well. We were all helping each other as a team and were looking out for each other. That is exactly what highly effective team players do, and we were very driven highly effective team players.

Other train operators in the UK were curious about what we were doing. They wanted to know why we had improved our fleet of trains by so much. We were getting visits from them down in Liverpool. So, I pulled together a few of the team and we presented our strategy to the other train operator's. They were really impressed and were going to try out some of the same ideas on their own fleet of trains. Which was a huge compliment to us.

As you can imagine, the low performing team players did a lot of moaning and groaning during this time. They were complaining that they had to do more work, and that the job was difficult. When in fact what we were doing was making their job easier. It just seemed harder because they didn't do a lot most of the time normally. Anyway, their complaints were falling on deaf ears. We knew what we did was great, and we were very proud.

Our engineering director Kevin began the new asset management strategy. It was his vision, his mission, but it was ours to deliver. He inspired us to take this on ourselves, as our own, and do something that made a huge difference to the whole organisation.

Kevin was one of the most driven people I had ever met, even now 10 years later. Even though he was the engineering director, he portrayed most, if not all the traits of a highly effective team player. When he went through the strategy with us as a team for the first time, he kept referring to ***"US"***, or ***"WE"***. Not "you", or "them". It was always *"We can make the difference."* Or *"There is only us who can make the difference."* It was phrases like this that inspired us as a team, especially me. I remember these exact words *"We will be the team who will have a positive influence on the whole organisation. We can make a huge impact and make a difference starting from now. We don't need to have a certain job title or be in a position of power. We can make a difference right now."* I will never forget those words. The reason I won't forget them is because I have used the same words when talking to my future teams. Kevin was a huge influence on me, and a role model. He was a very driven, highly effective leader. He was also a highly effective team player.

Are you driven enough to be the same or similar?

To be truly successful, we must be excellent. To be excellent we must compare ourselves against our own potential. Not comparing ourselves against others.

Have inspiration

A highly effective team player will never lose control of their emotions, beliefs, purpose, decisions, or their direction. Especially in terms of their performance within the team or as an individual. If they fail at something, they take full responsibility and learn from it. A low performing team player will blame someone or something for why they failed. They do not take responsibility for themselves.

To keep moving forward, and to consistently help to move your team forward, we must "have inspiration". Inspiration means that we are driving ourselves, we don't need anyone else to motivate us. It is internal. Motivation is when other people within our teams, our boss, or someone else that is driving us. It is external. When we need to be motivated, we are not inspired from within. An inspired team player still likes their boss to talk to them in a motivating way, but it is not essential. An inspired team player is going to move themselves, the team, and the organisation forward. They don't need any external motivation from anyone else.

A good example of this is you.

Did you buy this book because somebody motivated you to? Or did you buy this book because you were inspired from within to improve your leadership and teamwork skills?

If a person you know bought this book for you, they may be helping you through it with study. Or, they may have asked you to read it because it will help you. Either way, what the person is doing is trying to motivate you. If they motivated you to start reading this book, then hopefully from what you are reading, you are inspiring yourself.

However, if you didn't require anybody else to motivate you to buy this book, then you inspired yourself from within. You know in your heart that you need and want to become a better team player. There is a burning desire within you to improve, and as you work on improving yourself, you will also be helping others on your team to improve too. You will be motivating others within your team to follow you. It will be motivation to the rest of the team because you are external to them, but the internal desire will burn within them and will become inspiration.

You've come this far through the book. Are you feeling inspired from within, or are you feeling motivated by someone else? How would you know which one it is? If the book is interesting, you and you want to continue reading, then you are inspiring yourself from within. You don't require anybody else to motivate you to continue reading. However, if the reason you continue to read is because someone else is asking you to, then you are being motivated externally. That is the difference between motivation and inspiration. If we do something of our own accord and don't need other people, we are inspired. If we do something because someone else is asking/telling us to do it, then we are being motivated.

We are working on becoming a highly effective team player, but what does that mean? What is it we really want from being a highly effective team player? In my experience, we want to improve at our job, we want to be a better teammate, we want to have better teammates, and ultimately, we want to become an all-round better person. To be able to become all of these, we need to understand our potential and then letting it go from within us.

Now, let's understand that the reason we want to become a highly effective team player is not for more money. That is not our inspiration or motivation. We want to become a highly effective team player so we can help others, but also be happy and fulfilled within ourselves. We want to feel great about ourselves that we are making a difference. However, the benefits and rewards from becoming a highly effective team player will mean more money, more opportunities, and promotions. A highly effective team player is here to change things, and we do so because we are inspired, focussed, disciplined, flexible, and positive.

A low performing team player has the opposite focus. They are externally motivated by money, and all they really concentrate on is making more of it. I've heard it many times in the places that I've worked, *"I've been asked to work later, but I'm not getting more pay. I'm being used and abused."* A low performing team player has a very negative attitude towards going the extra mile for the team. They generally do want to make improvements, but they only want that so they can earn more money. They don't really care about feeling great about themselves, helping their teammates to improve, or being inspired from within. A low performing team player wants the rewards and benefits of more money, nothing else.

A highly effective leader does not want average team players. They want to have a team full of highly effective team players. There is nothing average about a highly effective team player. They are in high demand all over the world, and highly effective leaders are always searching for them. A highly effective team player is *excellent* not average.

When we are thinking about becoming a highly effective team player, and wanting to move from average to excellent, we must ensure we are thinking about the right things. It's very important that we must do the things that a highly effective team player does, not want what a highly effective team player has. Wanting what other people have will not get you what they have. We need to think about what they had to do to get what it is that you want. Wanting what a highly effective team player has is a path towards becoming a low performing team player.

A low performing team player resents a highly effective team player because of what they have, but they don't think about what they had to do to get it. If the low performing team player did what the highly effective team player did, then we would have a lot more highly effective team players in the world. So, make sure you think of the right things, and do what a highly effective team player does. Watch out for low performing team players trying to drag you down to their level and change your thinking.

When I was in my third year as an apprentice mechanical engineer in 2002, there were a lot of low performing team players around the plant. Most of them tried to bring me and the other apprentices down to their level while we were still new and inexperienced. On one shift, I was working on one of the Jaguar engine assembly lines, and our part of the line was to build the lift arms that would hold the engine block in place while it was cut into shape. I loved this job, so I would always get in early for my shift to prepare.

This didn't go down well with the other members of my team. One of them, Geoff, pulled me up over it. His words were, *"When you come in early Tom, you're making the rest of the team look bad."* I said, *"How is that Geoff?"* It was like something from the school playground. He replied, *"Well, as you know, we are all being watched closely because of fall in profit, and if you keep coming in early and looking better than the rest of us, it is going to show us in a bad light to the management."* I was baffled by this. The last thing I wanted to do was to make everyone look bad, especially when the company was low in profit, and we were being watched closely.

That is exactly the attitude of a low performing team player. They look at others to blame because they are not happy or not earning the money that they desire. I knew I wasn't making anybody look bad. There is only one person who can make you look bad, and that is you. What they should have done was, come in early like me, prepare and then the whole team would have looked good.

How do you feel when you go to work? How do your teammates feel? How does your leader feel? That is the most important thing when we are in the workplace, not what we do, but how we feel. That is what culture is.

Have confidence

When dealing with egos, sometimes the hardest one to deal with is the person who is looking back at you in the mirror.

Highly effective team players are also highly confident people. They are confident in their own ability to achieve great results. With being confident, they are also consistent in achieving great results which builds up momentum for them, the team, and the organisation. It is this momentum that attracts attention to the highly effective team player, and other people in the organisation want to follow them.

As we discussed earlier in this book, it is very important to have humility along with your confidence. There is the danger of becoming overconfident as you keep improving, and your results improve. Highly effective team players always have humility along with their confidence, this is a leadership trait that all *highly effective leaders* possess. I'm not saying you can't have pride in what you have achieved or feel great about yourself. But it is important that you portray an image of self-confidence and not arrogance in thinking how amazing you are.

How would this sound to you if I said this? "I am a person who believes in my own abilities. I am confident in myself. I take pride in what I do and what I have achieved in my life so far. I know who I am, and what I want to achieve in the future. I have a plan for my future." Does that seem arrogant or confident? It is confident. If you think that it seems arrogant, or you would find it difficult to have pride in yourself, then it is because of how you behave when you are taking pride in yourself.

Anyone who finds it difficult to be proud of who they are, what they've achieved, and their abilities are probably coming across a bit cocky, or overconfident. Confidence and arrogance are so close, and the only way to not appear arrogant or cocky is to have humility. To learn about humility then go back to chapter 1. If you don't have humility when you are being confident, then you will come across as cocky. That is a huge mistake when working towards becoming a highly effective team player.

A highly effective team player builds trust with his/her teammates, their leader, and the organisation. If you are cocky then then your teammates, leader, and eventually your organisation will not trust you at all. You will eventually be disliked by everyone. If that happens it will be almost impossible to become a highly effective team player. The main difference between a person who is confident and a person who is cocky is what they believe. A confident person believes that they are good at what they do. A cocky person believes that they are better than everybody else at what they do.

Highly effective team players use their confidence for the good of the team. They will use it to help raise the levels of the rest of the team and encourage them to become highly effective team players too. They use their confidence to keep adding value every day to their teammates and their organisation. A cocky team player will subtract value from their teammates and organisation. A highly effective team player will openly share their knowledge to help increase their influence, but most importantly increase the confidence in their teammates. A highly effective team player is always confident, and they are never cocky.

People who are confident are usually confident in two areas. They are confident in themselves through what they believe/know and experienced in their life. They are also confident in their abilities, skills, and competencies through what they have done either academically, or through their professional career. A highly effective team player is confident in both areas, and it's important that they maintain these confidences if they are to consistently keep achieving great results. It is also important if they want to help their teammates improve, and help their organisation grow.

Being confident in yourself means that you have core beliefs, and you live out your values 100%. Even when you are challenged on your core beliefs and values, you will not budge or be persuaded to believe something different. You do this because of the lessons you have learned, and the experiences you have had in your life. For example, my core belief is that leadership needs to be taught differently to team managers, team leaders, supervisors, and to the team. Hence why I am writing books and other materials on the subject. I know this because of my experience of being an engineer for the last 20 years, and an engineering manager for the last 10 years. Being confident in yourself is what makes you unique, and this confidence develops as you build your character.

Being confident in your abilities, skills, and competencies means that you are certain within yourself. You 100% believe that you can complete a certain task. For example, I 100% believe that I can teach and develop managers to become highly effective leaders, and I can teach and develop team members to increase their influence and become highly effective team players. If you keep developing your own competency, then this confidence will develop also. This is why we need to work on ourselves every day.

At present, you are learning something new in becoming a highly effective team player. So, you won't be 100% confident in your abilities, skills, and competence yet. So, you can't be 100% confident all the time. You may move positions from team member to manager, so you would lack some knowledge at first, and will need to build it up over time with experience and training. There will be times when you will lack confidence in other members of your team and their abilities to complete tasks. So, there are areas outside of your control that you will lack confidence in.

Being confident in yourself though is all up to you. At present you may or may not have self-confidence, but that is all about your character and if you are developing it or not. As you are reading this book, you are growing and character building. So, your self-confidence will be increasing. Your ability to learn new things is increasing too, and that again will increase your self-confidence.

You need to be confident in yourself if you want to increase your confidence in your abilities, skills, and competence. The two areas of confidence go hand in hand, and you need be confident in both areas if you are to become a highly effective team player. But you need to have humility in both areas too, otherwise you will come across as cocky and arrogant.

Highly effective team players are confident in both areas because they 100% know what they believe and what they know, and they are 100% confident in their abilities, skills, and competence. The reason they are 100% confident in these areas is because they are also 100% confident in what they don't know and can't yet do. They are confident in themselves to admit it. They don't try to "pull the wool over people's eyes" or pretend that they know something when they don't. They will go and find the knowledge they require from further learning or ask a member of their team who has the knowledge to help them. That is the only way they can successfully complete a task if they don't fully have the knowledge or skills and remain 100% confident.

Earlier in my career when I was in university, if someone was to challenge me about what knowledge I had, or skills, then I would get very defensive about it. Sometimes I would feel offended. Thinking back now, that was very cocky of me because I was not considering the other person's observation of me. They were not trying to insult me; they were trying to help me. I thought I was better than I was, and I couldn't see it.

Now, I look at this type of situation from both sides, and if the other person is right about me, then I am confident enough in myself about what I know, but also what I don't know. I take pride in myself that I have this confidence, and that is why I am sharing it with you.

I invite you to do the same in sharing the knowledge you have with your team, with others in your organisation, and even your friends and family. This will increase your influence, your confidence in yourself, and keep you on the right path to becoming a highly effective team player.

Accepting feedback on your abilities, skills, competence, and most importantly your performance is so important if you want to improve.

Make a difference, for the right reasons, at the right time

CHAPTER 5
MINDSET

Have courage

It takes character to put others before yourself and do the right things for the right reasons. It takes courage to make the difference, especially when nobody else around you is standing up to the plate.

In 2008, I was beginning to deliberately adopt the leadership traits that you are reading in this book, and in my other books: *"Manager to Leader – How to Be a Highly Effective Leader."* And *"Influence is Leadership – Become the Highly Effective Leader that others want to Follow."* One of the traits that I found difficult to adopt was "Have Courage." I think a lot of people find this trait difficult. As a highly effective team player courage is a trait that you must develop, and there will be times when you need courage in ways you never thought about before. For example, at times of change, highly effective team players take the lead from the front and change what is required. Leading by example takes courage, especially when you are leading without authority through changing times. I had to do this on numerous occasions, and I have worked with a small number of people who done this too when we felt the need for change. The reason for this was because our teammates were not doing anything to change things, even though they were unhappy.

Let me tell you a story to give you an example of this.

A friend of mine called Kevin worked with me in Liverpool for their train operating company. He was an outstation engineer, which means that when a train failed in service, he would attend the train to try and fix it and get it back into service. It was a demanding job because of the pressure from customers who were stranded on the platform and were delayed in getting to their destinations. Sometimes he was hurled abuse from some customers. There were even some dangerous situations, especially when the customers were football (soccer) fans. Anyway, Kevin had worked in the rail industry for about 20 years in 2010, so he had a lot of experience and knowledge.

The training for new and existing employees at our company was not the greatest in 2010, and there were a lot of complaints from people as you would expect. However, the people who were complaining had a lot of experience and knowledge too and did absolutely nothing to change this situation. The reason for that is because they were low performing team players. They blamed the management and leadership for the lack of training when they could have put a training package together themselves.

One day Kevin and I were having lunch and he asked me, *"How are you getting on with your degree and graduate training?"* I said, *"I'm getting on fine, but I need to learn more about the engineering and operation of the trains. I don't feel my knowledge is where it should be."* So, Kevin decided, *"Right, after lunch I am taking you onto a train and we are going to start on building up your knowledge."* So, straight after our lunch that is exactly what Kevin did. He did this for a couple of weeks until my knowledge was in a better place. I really thanked him for that, and I suggested to him that he would make a good technical trainer. And when I said that he told me, *"That is exactly what I want to do Tom."*

But there was no position available or vacancy for a technical trainer, or any kid of trainer in our company at this point. However, this was a burning desire within Kevin, and he wanted to change things to help all of our teammates within our depot. So, Kevin started doing to others what he did with me. He searched for people who needed to improve or refresh their knowledge on certain parts of the train, or in depot processes, and he taught them. He did this all off his own back, and people were improving. But, most importantly people were happier that someone was training them. He was getting no extra pay or anything, he just did this when he had spare time in his day. Or he would stay behind after work and teach. It was remarkable and very inspiring to see. It was also very brave, because some people may have got the wrong impression and got offended because they had more experience than Kevin. The management and leadership may not have liked it either, but Kevin didn't care. He had the courage to do this, because he had the passion and the desire to do it. Performance of the teams was improving, and the trains were staying longer in service without failing because they were being maintained better.

However, he had to keep this quiet at first because he wasn't a qualified trainer, and the managers didn't really know yet. Lots of the other engineers suggested to him that he should go to the managers and especially Kevin our director and tell them what he had been doing. But tell them with the recommendation that they make him the technical trainer. Now, this would take a lot of courage because he was doing this training without permission.

So, Kevin put together a document and presentation proposal which outlined what training he had been doing, why he was doing it (most important), and the results that the depot were having following his training. He then arranged a meeting with the depot manager and the engineering director so he could present to them his proposal.

Kevin had not really presented before in his career, especially to a senior manager, and a director. So, he practiced his presentation to me and a few other trusted teammates, and we all worked with him to make it the best it could be. He needed to make a few tweaks to his presentation and his delivery, but nothing major. This would help him with his confidence, and most of all help him maintain his courage to go for this. We were all behind him, and we believed in exactly what he believed. We believed that we worked for a great company, but we needed a new technical trainer and training team. We believed that Kevin was the man to do it.

It was a Tuesday morning, and this was the day of reckoning for Kevin to pitch his proposal to the depot manager Paul, and the engineering director Kevin. So, he went into their office, fired off his presentation to them confidently, and then sat down to be questioned. The first question was, *"Why have you been doing this, and why do you want to become technical trainer?"* Kevin answered, *"I first started this by training Tom Lawrence because he told me that he was lacking knowledge in the technical areas of the trains. So, I helped him. I have a passion for where I work and who I work with, and I want us to be the best we can be at what we do. The training facilities and trainers are not good enough at present, and I am the person to change that."*

Kevin was asked a few more questions, and then was told that they would think about his proposal and get back to him. Two days later, the depot manager Paul and engineering director Kevin got back to him and agreed that he should become technical trainer. Kevin was over the moon, he was ecstatic, and so was the rest of the depot. We were going to have the training department we wanted, and it was all down to our highly effective team player Kevin.

Kevin was a highly effective team player because he put his teammates and the depot before himself. He wanted to provide a platform for us so we can improve our knowledge, and most of all improve ourselves on being the best people we could be. I will always be grateful to Kevin for everything he did to help me.

When Kevin eventually took up the role as technical trainer, he created training packages and presentations for classroom training. These packages also included exams that had to be taken when the training package was completed. The company put him through college so he could gain the training qualifications he needed. Then he needed to recruit a couple of people to help him put the training facility together. Overall, it was a huge success, and he is still working as technical trainer today and loving it.

A highly effective team player understands that they can change things for the right reason, at the right time. They know that do that takes courage, and a highly effective team player knows how to build up that courage and implement. A highly effective team player doesn't just make the decision to make a difference, they make that difference. It takes an action not a decision to make a difference. It takes courage to make a difference, and it takes courage to be a highly effective team player.

If you think you have a weak leader, what are you going to do about it? This is your opportunity to take over and be the leader you wish you had.

Keep learning

Having the right mindset is the beginning of the success journey.

Regardless of a person's position, title, job description, or reputation within the organisation, a highly effective team player will always listen to them, with objective that they can learn something from them. Especially if the person is one of their own teammates. We can learn anything from anyone, no matter who they are. That is how a highly effective team player gets their insights. This is not just in the workplace either, they are willing to learn from anybody in all walks of life they come across.

Every member of your team will know something you don't know, and you know something that they don't know too. So, everybody has a wealth of knowledge that can help each teammate. The key is to have curiosity about what each of your teammates know, and what their outlook on certain situations are. Being curious is how we involve the whole team when there's a problem we need to solve. The closest team member to a certain problem, will know more about the problem than anyone else. So, they will be able to give us the most information about it that will help solve the problem. A highly effective team player will listen to them first, and then involve the rest of the team when they need to find out more information that will improve things.

Highly effective team players are educated through school, college, maybe university, just like everyone else does. However, they don't stop their education there. They believe that they need a lifetime education that will span 60 to 70 years, and they need to keep learning every single day of their lives. If they don't do this then they believe that they are missing out on something, and it will irritate them. Personal growth is constant with a highly effective team player, and never stops. The more a highly effective team player learns, the more aware they are of the things in their life that they don't know about. So, the more a highly effective team player knows, the more they realise what they don't know.

A low performing team player relies on what they have learned either in school, college, university or from their last jobs. They don't believe that they should continue educating themselves. For a low performing team player, education stopped when they no longer had to visit the classrooms and listen to teachers or professors.

The average person believes that an education is school, college, or university. They believe that when school, college, or university is over that their education is over. However, this is not true. To be a real student, and to be a highly effective team player, you must 'educate yourself' every day of your life. All a school, college, or university can do for a person is to lay the foundations for an education. It is then up to the student to work on themselves every day to keep learning. A student loan, or a sponsorship from an organisation or industry cannot get you educated. It is just the building blocks to start a person off on their daily learning journey. An education requires hard work, and it is not completed after the usual four years of a college or university. It is a lifelong journey, and the more you realise that now, the better it will be for you. Highly effective team players are aware of this, and they work on themselves, and educate themselves everyday by constantly learning. No matter what the source of learning is. It can be from their leader, a teammate, a book, a website, a family member, it doesn't matter. All that is important is that you are constantly learning and learning the right things.

We have spoken at length about having humility as one the important traits that a highly effective team player must have. When you are constantly learning, having humility will keep you grounded, and will allow you to have the right attitude to learning. Humility is a thread that goes through most of the traits required to be a highly effective team player, especially when you are learning something new. Without humility, it won't be possible for you to become a highly effective team player.

Low performing players who have an attitude of, *"I don't need to learn anything new; I know everything I need to know already"* are arrogant and have a huge ego. The things that they don't know, they will try to disguise to people to make them think that they do know those things. That is what they want everyone to think. A low performing team player will not listen to anyone else who is trying to share information about a problem, or something important. They think that listening to other teammates, or others within the organisation is a waste of their time. They already know everything, so why would they listen to others right? I have worked with many low performing team players who had this attitude, and I've even experienced them saying to me that I didn't know what I was talking about.

When I was working in Scotland as a production manager, I was advising one of the other production managers about an issue with a train engine. He was more experienced than I was, but I knew more about the problem with the engine than he did. I told him, *"We need to isolate this engine because it had a chance of failing when going back into service."* He replied, *"Tom, you don't know what you're talking about. The engine will be fine, I am not having it isolated, or otherwise we will get a kicking from the boss."* So, I let him get on with it. I could have argued the case, but I let him make his decision. What happened next? The engine failed in service. So, the lesson there is, listen to the person who has the most information about the problem.

The reason I believe that as a highly effective team player, we must be learning every day is, it allows our teammates, and others to be responsible in helping you. By them helping you, you are helping them increase their confidence in sharing information and communicating with others. Especially if what they are sharing helps you and the team in becoming successful or achieving a great result. So, it is a win/win situation for both, as you're both teaching, and you're both learning at the same time. You may already know how to help your teammate build their confidence, or a new teammate with no experience. But, allowing them to teach is the perfect way to allow them to feel responsible, with in turn will build confidence in them.

Another reason is by allowing your teammates to teach you or share information with you will help you build a relationship with them. A great trait for a highly effective team player to practice is listening. Even if you know what your teammate is talking about, still listening to them will make them feel good. Which will help you both in building your relationship with each other. When building a relationship, you are building trust. So, by building trust, you are both more likely to share information, and even more important information with each other in the future. If you do not listen to your teammate then that will create distrust and will be disrespectful which will break the relationship.

Highly effective team players have a personal mission to be a role model for their teammates, and part of that is to help the team develop and grow. When their teammates are teaching and sharing information with them, they are being humble and role modelling other leadership traits. When the team sees this, they want to do the same and follow the behaviours of the highly effective team player. When this happens, the team are all becoming highly effective team players. It is a snowball effect because the other teams in the engineering department will also see this and want to model them in the same way. Highly effective team players always remember that it starts with them, but it is not about them.

A highly effective team player will never question themselves of whether they are a role model or not. They know they are a role model because they know they are being watched by almost everybody. The real question is, *"What kind of role model am I?"*

So……...***what kind of role model are you?***

By knowing how much we don't know, it will give us the urge to find out.

Have an open mind

Being open minded keeps us curious. When we are curious, we never stop learning new things, or finding new ways. Curiosity builds momentum, so never break the momentum.

A highly effective team player loves to ask questions, especially of their own teammates. They do this to prompt their teammate to think deeper than they normally would. So, the type of question they would ask would be a probing or a coaching type of question. This is also an opportunity for the highly effective team player to learn, as we have just discussed. When there is a problem, they question their teammates so that they can come to a solution as a team, not as an individual. This requires the team to share their ideas, and not be satisfied until they come to the best solution possible. Not just any old solution.

A low performing team player will go out their way to find a solution to a problem on their own. They do this so that they look the best compared to everyone else in the team. They crave the attention of their line manager and others in their department. They see this to compete with everyone else, rather than work together and collaborate as a team. By providing a solution on their own, they are expecting to get all the credit for it. This is a very close-minded way to work when you are part of a team. They can't stand to have another team member provide a solution to the same problem, because it will take away any credit that they wanted. So, it ends up not being a team at all, it is just a group of individuals.

When it comes to highly effective team players, there is no comparison between anyone. Everyone is on a level playing field. Everyone's ideas, opinions, and solutions will always be listened to and considered. A highly effective team player is not searching to get credit from anyone. However, they do get a lot of credit for when the team achieves a great result. The great thing is though, they pass credit on to the rest of the team and share it. As they are humble, they share everything they get with their teammates, so that everyone feels part of the achievement. That is how a highly effective team player raises their team up.

When I was in my early 20s and going through university whilst working for Liverpool's train operating company, I was intent on being an autonomous person. Especially when it came to my work, as I thought that being autonomous was a sign of strength to my teammates, and to my boss. I was making progress on what I was working on while increasing my influence when I could, and I was continuing to build relationships throughout the organisation. If I needed help from somebody, I wasn't afraid to ask.

But that was where I ended it. I resisted any offer of help from people because I was of the frame of mind that accepting help from someone was a sign of weakness. Why would I need help from someone if I was continuing to make improvements each day, make progress in my work, and feeling good about myself? I couldn't allow myself to show weakness to my teammates or my boss, otherwise they might think differently of me. When, not accepting the help and thinking that it was a sign of weakness, was a sign of weakness. But, at the time I thought it was a sign of strength.

It wasn't until I had my graduate mentoring session with Ron that we discussed my progress that I started to think differently. I was advising Ron on the work I had been doing and improving the performance of the trains, and he seemed happy. He then asked me, *"How has Tom been doing and feeling since we last met?"* I said, *"I've been fine, and I've been working autonomously like you told me to."* He replied, *"That's good, have your team been helping you too?"* I was confused, *"I thought autonomous meant you don't need the team. I thought this was my time to show what I could do."* Ron's answer to that was, *"Absolutely not Tom. That is not what I meant. You are working on becoming a leader, leaders need their team to help them. If you didn't have teammates, then you wouldn't be able to lead."*

When my session with Ron finished, I was so disappointed in myself that I had gotten completely the wrong idea to what Ron had advised me. He told me to be autonomous, meaning I can work alone when it was necessary. But it doesn't mean you can neglect your team or refuse to receive their help when it was offered.

Over the next few days, I thought about Ron's words again and again. Instead of feeling like I was showing strength by working autonomously, I now felt bad and weak. Why would I think that I can do something better on my own, instead of working with the help of my teammates? Why would I think that I don't need my team? It was eating me up. I was trying to be positive through being autonomous, and when somebody offered me help, I rejected it in a nice way, so I wasn't being nasty to anyone. But I wasn't thinking about how I made others feel. They might not have shown it, but I'm pretty sure my rejection of help would have hit a nerve with the teammate who was offering.

I wanted to get to the bottom of why I was thinking in this way. Why did I not accept any help from my teammates? Why did I think I could do things better on my own? Why did I neglect my teammates? Why had I stopped listening to people? I spent quite some time thinking about these questions and I was determined to get to the answers.

This reflection on these questions carried on when I went home from work. They were constantly in my thoughts. So, I asked my dad what he thought. Straight away he said, *"You are being closed minded Tom. You're rejecting things from people because you are seeing things very narrowly. You need to be more open minded."*

That was it. My Dad was so right. I was behaving in a closed-minded manner. The same way a low performing team player behaves. It was like I was competing with my other teammates on who shows the most strength. If I was the most autonomous out of everyone, then I would think I was the strongest member of the team. When, I was the weakest by not working together with my teammates.

I had gotten to the bottom of it, I was being close minded. I decided then that I would be open minded in the future and always include my teammates. Especially when it came to making decisions or coming to solutions to problems. Being closed minded to your teammates gives them a very negative impression, even if you think, you're being positive. Not listening to other people's ideas because you don't think you need them is also very negative and can create distrust within the team. So, we must be open minded if we want to continue building relationships.

What I had done was a huge mistake on my part, and I needed to admit that to the team. I needed to stop giving people the impression that I was a negative person too. So, being open minded was one thing, but I had to turn it around. I needed to come across in a positive way when somebody approached me. How I figured out how to do that was by accepting help, listening to ideas, and sharing my own ideas with the whole team. It all comes down to sharing your thoughts, ideas, and information with your team.

A team of brains is better than one brain. So, keep that in mind when you are at your next team meeting.

Learning from mistakes is the only way you will become successful, no matter what it is you are doing.

It's not what you know, it's not who you know…it's both

CHAPTER 6
SERVICE

Connect with others

Having the ability to communicate, and most of all connect is first and foremost how you become successful. If you can't do that, then you need to learn how.

Now that we have started chapter 6, we are 50% of our way through this book. Before you carry on reading, have a think about what you have read so far. How are you going to implement what you have learned so far, and what you will learn in the rest of this book? How are you going to help the others within your team, to work towards becoming a highly effective team player? How are you going to help the other teams within your department, and within your organisation?

Using your imagination, what would your team and your organisation look like with a lot more highly effective team players? Imagine the low performing team players changing their attitude and making the decision to work on themselves with your help. Imagine how you could help your leader, and the other leaders too. What could you do to help and support them? How could you help them to turn their teams into highly effective team players?

You have a new wealth of knowledge from this book, and other resources. Imagine the difference you could make in your organisation, and your industry with your new knowledge.

Thinking about your own team first, who on your team are highly effective team players? Who on your team are low performing team players (if any)?

When you go back to your team, what are going to discuss with them? What action will you take to help improve your team, both as a whole and as individuals? How are you going to make your organisation a great place to work? How can you help your organisation bring in new and better talent? How can you help your organisation retain their highly effective team players? How are you going to feel when you go back to work, and you are with your team? How are they going to feel? How can you help your teammates to start feeling better about themselves, and their place of work?

These are all very direct questions, and you don't have to answer them all at once. But they are questions that you should think about while you are reading the rest of this book. You should also think about how to answer these questions when sharing the ideas from this book with your teammates.

Everybody can improve, not least the low performing team players that we have on our own teams, and who are on other teams within our organisations. However, for us to continuously improve, we must work on ourselves every day, and to do that, we must have the right attitude. As we have discussed on many occasions, a low performing team player's attitude to change or improvement is wrong. So, when sharing these new actions from this book with low performing team players, we need to work on helping them change their attitude first. We do that by making a connection with them.

To make a connection, we must first show our low performing team players the way by leading by example. We must commit to not only learning the actions that are on these pages, but by deliberately choosing to implement them for ourselves first. We need to prove that these actions work, and when doing so, we will start to see attitudes changing.

Our influence will increase with our teammates, and the relationships that we have will become stronger. With increasing our influence with the low performing team players, we will start to build trust and relationships with them too. Things will start to feel better within our own teams, and throughout the organisation.

Making the choice to lead by example is one that we must make when we want to implement what we are learning in this book. We cannot let others make any choices for us, and we cannot let what we are learning go to waste by doing nothing. So, let's make the commitment to action implementation today, for ourselves, for our teams, and for our organisation. Let's choose to be the leader. It's up to you.

The first time I made the choice to take action with my team and help them to work towards becoming highly effective team players was when I first moved to London. I worked as an engineering performance manager for Siemens, and I was part of a team based at a west London train depot.

One of the team was a technical engineer called Phil. He was a good engineer. However, he wasn't a team player, and wasn't very approachable when it came to teamwork, or other team-based activities. At the time I considered him a low performing team player because of his lack of consideration for the team. All he really cared about was getting his work done, making sure the trains were working well, and then going home at the end of the day. He made sure the bosses knew what he did, but not what the team did.

There was one occasion in 2015 when I had just employed a new technical engineer to join Phil as part of the engineering performance team. His name was Freddie, and I wanted him to work closely with Phil at the beginning for two reasons. The first reason was to try and bring Phil out of his shell, and embrace that he was part of a team, and that his job was to help his teammates. The second was to see if Freddie could learn anything from Phil, because Phil was a good engineer. Neither reason worked.

Freddie came back to me and told me that Phil was not working well or co-operating with him at all. Freddie asked Phil, *"How have you managed to keep the trains performing so well for so long?"* Phil's reply was, *"I'm sorry Freddie you will have to find out for yourself. I am not willing to give away my own secrets on how I do my job well. Nobody helped me, so I am not going to help you."* He was not an approachable person whatsoever, and he was making life difficult for everybody.

Phil was not connecting with anyone, including me, so I needed to take action. I sat Phil down for a one to one, and I asked him the question, *"Would you consider yourself a team player?"* He replied, *"Yes."* I asked, *"How?"* He said, *"Because I am the one who is keeping our trains performing, nobody else."* I couldn't believe what I was hearing.

I continued, *"So, how is the fact that you claim to keep the trains performing, and nobody else, a part of teamwork? What was the team's contribution?"* He couldn't answer that question. He continued to tell me what he had done, and what he was still doing to keep the trains running.

Phil's problem wasn't that he was an ignorant person, or even that he had a bad attitude. It was the way he had been trained throughout his career. He was taught that it was every man/woman for themselves. From his previous places of work, teamwork was not really promoted or supported. So, I decided that the action I would take was to teach my own team about what a team is, what a team does, and how we can make things better by being a team. This was my opportunity to lead by example with my team, and hopefully the rest of the department.

We spent two days in the classroom, and I taught my team "what is a team?" "How can we be a team?" "What are my expectations of us as a team?" Then finally, for the team to come back to me on how we were going to achieve my expectations. All I did was listen, I left it up to the team to work out how they would gel as a team, and how they can achieve the desired results.

Listening to the team speak to each other with such passion, and their plan to achieve my expectations was the most enjoyable part of the two days. I knew it was going to work because of the vigor in each person's voice. The passion I was building for the team felt amazing.

They connected with each other, and with me. I had taught them how to be a team, and they were now implementing what they had learned. Over the next weeks and months, Phil had a transformation like none I had ever seen. He was becoming a highly effective team player, and a leader. He was openly sharing his ideas, his knowledge, and even helped apprentices and graduates when they visited the depot. I was and I still am extremely proud of Phil, as he still works in the same team and is helping the Siemens organisation every day.

To connect with others, you must build relationships with them, and to do that you must share what you know with them. That is how you build trust. When you connect, you increase your influence. First within your own team. Then multiply your influence with your department, your organisation, and your industry by continuing to connect with others.

To build relationships with your team, you need to spend time with them. You need to share your knowledge, your ideas, your opinions, and you need to help them.

Help others

Showing humility every day is not easy. The best way to do that is to help others and live a life of servanthood.

Helping others is a choice that we all must make. We either want to help others, or we don't. The day I decided to start helping others within my team, was the day when things started to change. We started to perform better as a team, we all started to help each other, and our relationships were getting stronger. When we decide to start helping others, we are deliberately building relationships, and trust with the person we are helping. When we as a team started to perform better, our department and our organisation was performing better too. It has a multiplying effect across everyone when we all decide to help each other.

If it wasn't for all the personal growth activities, I have done over the last 20 years of my career, I wouldn't be able to write this book. I am so proud to be able to teach you these lessons from my own personal experiences. All the leadership books I have read, videos I've watched, events I've attended, and work experiences are all rolled up into the resources I provide. There are more resources to come too. This is all help for you, and the others who you share this book with. The many mistakes I have made in the last 20 years, the lessons that I have learned from those mistakes, and the person I have become, is my way of helping you.

Now it is the present day, and the person I have become is a highly effective team player, and leader. However, nobody is perfect, so I am constantly maintaining myself, and overhauling myself. I have teams and mentees that I need to help, and I am responsible for. I have a purpose and I am living and breathing that purpose every day. It is my job to help you to be the same, whatever your purpose is. It is my job to help you to become the person you want to be. It is my job to help you become a highly effective team player and leader.

As I've explained earlier, I once had the attitude of every person for themselves when it came to my work. I thought that if you could do tasks on your own, rather than accept help from teammates, it showed strength. When, it wasn't. So, when I realised this, and I got over myself, I started to help others. This started in 2009. Since then, I have made it a priority to always help others. No matter if they were in my team, from another organisation, or industry. It has been a 10-year journey of being helpful, which coincides in exactly the same amount of time I have been studying and working in leadership. I couldn't have imagined back then, the teams I would be a part of, and the size of the teams I led. Back in 2009, I didn't have the vision or the size of team that I have now.

As you and many other people are reading this book, and my other books, I am on yours and every other reader's team. I want to help you. I want to help the people who watch my video's, listen to my audios, read my blog posts, and read my social media posts. My followers are ever growing, and I want to help every single one of you. You are very important to me, and anything I can do to help you become the person you want to be, and a highly effective team player, I am going to do.

I have a purpose and a burning desire to help more and more people every day. I want to do this while I am also working on myself to become the best person, I can be every day. However, I put others first, as that is how I have learned from others who helped me in the past. Hopefully what you are learning will enable you to do the same when helping others. I want you, and the people you help in the future to become great. I want you to become highly effective. I want you to become a leader who can make a difference in the world and help change the world for the better.

A lot of people who do the same thing I am doing, writing books, and creating other resources, do so for the money. That is fine, I have no issue with that, except they are telling their customers that they are doing it to "help others." When, their objective is to earn money. For me, money is important, but it is not the most important thing. The most important thing is that I help you to improve, and become a highly effective team player, and a leader. The fact that I get paid well for it is a by-product of fulfilling my purpose. The more people you help from applying what you learn in my book, you will eventually have the same result as a by-product too. However, if you focus on money rather than helping people, you will not give absolutely 100% to your cause and putting others first.

No matter what team you're on now or will be in the future, there is always somebody you can help. Whether that be help in the task they are currently doing, help in their job, or help outside of work. If you can help them, then it is an opportunity for you to increase your influence and build on your relationship with the person. When I started to deliberately help people, I didn't care what they needed help with. If I was able to help someone, then I would do it. If they needed any knowledge that I had, or needed my help to find knowledge, or needed me to check their work. Whatever it was, I was happy to help, because I had a newly found purpose that I had to fulfil every day.

Most people on most teams throughout the world are not taught that helping each other is the best way to produce great results. I was always taught to get on with the job, and if you had a problem then tell the boss. I have interacted with a lot of different colleagues over the years, and most of them have been taught the same thing. So, these team members will not realise how valuable they are to each other, and that just a little bit of help to their fellow teammate can be of great value. As they are not aware to make helping others a priority, they are falling short of fulfilling their potential. Each team member has unique experiences that they could share with others and help them with. If they do not do this then they are holding themselves, and their teammates back. It is our challenge to change that and turn these team members into highly effective team players. Are you up for that challenge?

You have the potential to create so much value in the world by helping others. By helping others and sharing your knowledge and ideas, you are fulfilling your potential. Don't allow yourself or anybody else to hold you back from that. I believe you can do it. Do you believe you can do it and help others?

Believe me, when you start on your journey to helping other people, it will take many paths. What I mean is, you will be helping people on your team. You will be helping friends and family with any issues that you can contribute to. You will be helping people in the local pub. You will be helping people down your local golf club. It is a bug that you will not want to get rid of, because you know you are serving your purpose and you are fulfilling your potential. Every day I wake up, I leave the house knowing I need to help somebody today, otherwise it will not be a productive day for me. You will start to feel the same. Trust me.

A highly effective team player is not the most valuable member of the team, he is the member of the team who makes the most players valuable. That is how a highly effective team player helps and lifts their team. You will also begin to do this as you start to share what you are learning on these pages. The impact you will have on your team members, and the others that you will help will be very powerful.

Helping others is very important when it comes to being a highly effective team player. The sense of purpose it gives you is so high that you will want to serve it every day. You will still have enough strength and energy to ensure you work on yourself too. The difference you can make in the world by helping others is huge, and it should not be underestimated. How big of a difference do you want to make?

You and I have the ability and the resources to help a lot of people. It takes courage to start. Who is going to be the first person you help?

Have empathy

Having empathy is seeking out opportunities to help others. Think of yourself less and think of others more.

If there is one thing that I have learned over the years when I want to help a person, in any capacity, it is to "Have Empathy." What does that mean? It means to seek out opportunities to help others. How do you do that? Well, you think more about others than you think about yourself. That doesn't mean to put yourself down or think of less of yourself. It means to think about yourself less, but still love yourself. When I say love yourself, I don't mean that in a cocky or arrogant way. I mean value and respect yourself.

When a highly effective team player is surrounded by, and working with their team, they think about the team more than they think about themselves. With that attitude, they are set up to seek out opportunities to help another teammate. They still take care of themselves, and love themselves, they just don't think about themselves as much as they do their teammates. The needs of the team are important than the needs of the individual. However, it doesn't mean that the needs of the individual are wiped out completely. They still count. You must still value yourself, just be aware of the team's needs and their values a bit more.

During my apprenticeship I worked with many talented engineers, but there was one that stood out for me. His name was Tony. He stands out, not only because he was an excellent engineer, but because he showed a lot of empathy. He was about 10 years older than me, so he had a lot more experience, and knew the job a lot better. We worked on building the car engine assembly lines together, and he helped me lots, even when I didn't know he was helping me. He had this way of making you feel like everything was your idea, even if it was his idea really. His objective was to make others feel good.

Tony was very deliberate about being empathetic with his teammates. He knew that by helping others, the team would work together and collaborate a lot better. He wanted us to beat our expectations, and he felt the only way to do that was by helping the team. When Tony came to work, his focus was to use his own experiences, and skills to help us. Even when it came to opening doors for people or picking a tool or something up for another person, Tony was right there. Every day he would give me a lift home from work because in those days I couldn't drive. I remember talking about Tony to my dad, saying I wanted to model him as an engineer, but also think of others like he did.

Tony loves to help others and think of himself less because that is what was instilled in him. He has the heart to think of others and help them where he could. In the words of someone from Liverpool, *"He was a lovely lad."* You might be thinking now that you don't have any empathy or think of others as much as Tony does. But don't worry. Empathy can be learned. I had to learn it. You may have gathered that; due to the stories I've told earlier.

I remember walking past the notice board at our place of work, and I saw a notice that Tony had put up looking for people to play football (soccer) on a Wednesday evening. He needed 12 players as it was 5 a side, and each team would have a substitute. I was first to volunteer because I loved playing 5 a side, but most of all because Tony was organising it. Within a day or two, we had all the players we needed, so the next Wednesday we were playing 5 a side football. It was a great way to keep fit, and it was great for socialising outside of work. That was what Tony put on the notice board. He was thinking of helping people keep fit, and it was good for team building and getting to know each other better. I wasn't the fittest at the time, so I played in goal for the beginning. Tony would always replace me because he knew I could get fitter by playing. He was thinking of my fitness more than his own. He did this with all the players too, not just me.

So, if you're thinking of ideas for team building exercises then team sports is an excellent way to start. A lot of teams go for nights out to the pub or a bar. This is great, but if you're doing activities where you need to work as a team, then sports are much better. When we played 5 a side, it is a very fast sport, and we needed to help each other to score a goal or defend. Then afterwards, we had the social side where we had a lot to talk about. We talked about the football, and how we can improve things at work. That was Tony's thoughts when starting the 5 a side. Fitness, team building, and brainstorming. Again, he was being empathetic to our needs as a team. He knew that to improve things we needed to brainstorm with each other, and the best place to do that was outside of the working environment.

One idea that we wanted to pursue was, instead of working individually when building the engine assembly lines, why not try working in twos. Tony asked us what we thought of it, and as part of the brainstorming we drew up all the pros and cons. There were more pros than cons when we were finished, so we made the decision as a team to trial it. Well, it worked. The guy I paired up with was called Steve, he was about 4 years older than me. So, he still had a lot to learn, but Tony thought that two younger heads working together might work well. Guess what? We worked well together. The best thing about working in a two was the help we gave each other. That was Tony's thinking when he thought of working in pairs. He knew it would force us to help each other and help move the team forward.

For the rest of my time during my apprenticeship, and the few months following, I always worked in a pair when building an assembly line. All down to Tony's initial idea, and then the team's brainstorming of that idea. So, to this day, I always encourage colleagues to work in two's whenever it was practical. Sometimes we must work on our own, but when we are surrounded by the team, working in twos was my philosophy. Most of the time it worked. But there were times when people didn't get a long, or they preferred to work on their own. You might be one of those people now, but believe me, working alongside another can only be good for the team overall.

I bring you back to the story of Phil when we worked in London for Siemens. I wanted him to help our new team member Freddie by working with him and sharing his knowledge with him. He didn't want to do it at first, but over time he came round to it with a little encouragement and leadership. If you have any team members like that, then all you need to do is share what you are learning in this book, and they will come round to the idea of working alongside their teammates and helping each other. It is the only way the team can achieve great things.

I wouldn't have thought like this if it weren't for my mentor Ron, but also having the role model of Tony in my apprenticeship. I changed my thinking a bit when I got older because of the people I was working with, and that I thought receiving help was a weakness. But my mentor gave the lessons, and that is when I started to think of Tony again. Unfortunately, I only got to work with Tony for 4½ years because the company we worked for went bankrupt, and we were all made redundant. Following that, Tony set up business for himself and I went elsewhere, but I will never forget him. He had a lot of empathy, and because of that he was a highly effective team player. Everyone in our team trusted him. He was a very valuable team player, and he cared more in making us more valuable. That is what empathy is all about. Using your own skills and talents to benefit others, not just benefit yourself.

Having empathy and helping your teammates is what builds the relationship between everyone and builds trust. You will create distrust if you don't have empathy or want to help anyone. How are you feeling now about empathy and helping others?

Increasing your own influence is what leadership is. Helping someone to increase their own influence will multiply your influence.

If you involve people, they will engage. If you leave them out, they will disengage

CHAPTER 7
ENGAGEMENT

Engaging with others

What is your responsibility? YOU. It is your responsibility to ensure that you have everything you need to get a job done. If you need tools, get them. If you need information, find it, or ask someone. Do not wait for tools and information to just come to you, and then blame others for not getting them. That is the act of a disengaged team player, not a highly effective team player.

Taking ownership of the team, and the results that the team produce is what a highly effective team player does. He/she will engage with others in the team and be engaged enough to go the extra mile. In other words, they do more than the expectations of their team leader, or the expectations of the organisation. A highly effective team player will not rely on anybody else, either within their team, or from outside their team to take responsibility for the team's results. If they did, they would not be highly effective.

A highly effective team player will always share the information they have that the team need. They will not guard it or wait for one of their teammates to come and ask them for it. They are engaged enough to know what the team needs, and then will engage with the team by providing what the team needs before they need it. If a highly effective team player needs information themselves, they will engage with the people who have this information and ask them for it. They will not wait for someone to approach them to give them the information, and most of all they will not blame others for not providing the information they require. A highly effective team player is very deliberate when seeking out information, and by engaging with the right people, they will get what they need.

In the team you are currently in, who are the most engaged teammates? What level of engagement do you have with your teammates, and your organisation? If your leader had a team of highly effective, fully engaged team players, imagine what you could achieve.

When a low performing team player has information that their team needs, they prefer to hold onto it and guard it. They do this because they think that by providing others with information that will help the team, will help others get ahead of them. It is a very paranoid mindset that they have in these cases. Even if a teammate asks them for this information, it will be a strain for them to give it to them. However, when a low performing team player needs information that will help them, they expect a teammate to approach them to give it to them. They will not seek out the information for themselves. By doing this, it gives them an excuse to blame others for not providing them with the information they required. They think by blaming others, it takes the bad light off them and shines it onto the rest of the team. However, most people can see through their excuses. This is an extremely disengaged attitude of a low performing team player.

I have first-hand experience of being part of teams and leading teams of disengaged team members with poor behaviour. For example, you will hear low performing team players saying things like, *"Nobody told me, so how am I supposed to know?" "Didn't you know that I needed to know that?" "You should have told me earlier, not today." "I'm sorry, but that's not my responsibility."*

You have probably heard similar statements to these yourself. You may have even said these yourself. I know I used to say them before I started to work on and develop myself.

One thing I have realised over the years, especially in the engineering industry, is that people are very reluctant to go the extra mile unless they are paid to. I have worked in teams that worked shifts. By working shifts, it meant you would get paid overtime if you did extra hours. So, whenever we were asked to stay on extra hours, we would gladly do it because we got paid time and a half, or sometimes double time. However, when we were taken off shift, and moved to just salary-based hours, it meant there was no more overtime. So, whenever we were asked to do extra hours, we ALWAYS said, *"No way"* because we were not being paid for it. That is a low performing team, and an extremely disengaged mindset to have.

If you hear one of your teammates say things like, *"I'm only here to get paid."* Or *"I'm not doing that, it's not my job or in my job description."* Then you will know they are low performing team players and are disengaged with what the team is trying to achieve. It's very sad to hear people who are on your team say things like that. They are basically admitting that they are happy to stand back and let the rest of the team do the tasks that they know they can perform. They are admitting that they don't care if the team achieves good results or not. They do not care about anybody else but themselves. Are you ok with having people like that on your team?

I ask that question because I am not ok with people on my team behaving this way. By being disengaged with the team, and what the organisation is trying to achieve, they are not helping anyone. Especially themselves. A low performing team player is never going to be offered any new opportunities, pay rises, or job promotions. They think the reason they are not being offered these things is because of other people. So, they blame others for their own shortcomings, which is completely wrong. The fact that they are still in their job, after several years of behaving in such a way is a miracle in my eyes.

As a highly effective team player it is your responsibility try to help the disengaged become more engaged into what the team is trying to achieve. You should also help the team engage into why the team is trying to achieve it, and ultimately what the organisation is trying to achieve. So, how do you do that? Well, we must try to understand why the disengaged person feels and behaves in such a way. What is it that is driving them to be like that? Why are they not engaging like we are as highly effective team players?

Sit down with them and have that conversation with them. As a highly effective team player, you are seen as a leader, so it is your job to listen to the person to understand what it is. Then react to what you now understand with some actions that you are going to take to help them. Remember, as we spoke about earlier, it is about ***HELPING*** them to become more engaged. That could be taking them under your wing and leading by example. It could also be by explaining why you are an engaged person, and why you want to achieve great results. Doing whatever you can to help them change their attitude is what counts.

I was once a disengaged person, and I used to blame anyone but myself for things that didn't happen in my life, not only at work either. I was focussed on blame, and I did not take ownership for anything. I believed that anything negative that happened in my life was someone else's fault. That was until I started working on myself every day and took ownership for my life. After all, it is ***MY LIFE***. That is what a low performing and disengaged team player needs to realise. We need to help them realise that, and then offer more help until they can become engaged. I received help from my teammates and my mentor Ron who you now know about. This is your opportunity to do the same with any negative, or disengaged people you know or work with.

When you can help a person to become more engaged, you will start to see differences in that person. They will stop blaming others, especially teammates. But, most of all, they will become more of a positive person, and will try to help others to become more engaged by following your example. It is a snowball effect, and it is very addictive. I say that because when you see the differences in the person you have helped, you want to do it again, and again.

Being a positive, highly effective team player is how it starts. Turning a negative attitude into a positive attitude is how it begins with becoming more engaged. If you can do that first, the rest is easy.

When you help a person change, just watch to see how much they want to change things for the better. The person who would gladly stand back and watch the rest of the team do the work, is now on the front lines trying to be involved in everything. The person who would not go the extra mile because they were not getting paid, is now going miles and miles whether they get paid or not.

The feeling when you see that change in a person because of your help is amazing. Do you want that feeling, and do want to see that change in your team?

A highly effective team player is a very productive person. Productive people are looked upon as leaders. How productive are you? Would you consider yourself a leader?

Constantly search

If you attend a 3-day training course and you come away with one great idea that will help you and your team, then the 3 days has been worth it.

Highly effective team players are constantly searching. What are they searching for? You might ask. They are constantly searching for things that will increase their influence, increase their personal growth, and increase their value to others. A highly effective team player knows that for them and the team to become successful, he/she needs to become more of a valuable teammate. They have an individual purpose, and they live that purpose every day. Every task they carry out either in work, or outside of work is done on purpose. Everything they say is said on purpose. They live with purpose by constantly searching for things that will make them more valuable.

I have led engineering teams across the United Kingdom in various departments. In Liverpool I led a project team. In Edinburgh and Glasgow,

I led production teams. In London I have led a performance team, a technical team, and another project team. I also mentor young engineers, and have given talks to in Liverpool, Glasgow, Utrecht in the Netherlands, and London. Some of the talks I gave were cross industry events. The industries present included the rail industry, the automotive industry, the aerospace industry, and the oil and gas industry. The reason for that is to see what each industry could learn from each other to improve.

With leading all the different teams, in different parts of the United Kingdom, working with young engineers, and giving talks to different industries, there is one thread that goes through all of them. I share one message to ALL the different teams of people, and that message is, *"Constantly Search."* What I mean when I deliver that message is search for things that help you increase your influence, increase your personal growth, and make you a more valuable person. What helps me to constantly search is; read personal growth or leadership books, watch personal growth or leadership videos on You Tube, or attend a conference/event. That would be a good start for you to start your search on becoming a more valuable person. However, don't try to take in everything you read, listen to, or watch. Only write down or remember information that will help you to become more valuable. Some of the information you consume will not help you, so get rid of it. Valuable information is what you are searching for, nothing else.

I mentioned that I have spoken to engineers from the oil and gas industry. For oil to be found, it needs to be searched for. Most oil is buried underground, and oil companies' use drills to break through the rock to get it. Oil rigs search for oil out from under the seabed. A lot of the time they find oil, but the rest of the time they don't. However, it doesn't stop them from searching for it. They keep going until they find what they want, which is the oil. Oil is what will make the organisation they are working with more valuable and will help the people they sell it to.

A highly effective team player must use the same philosophy when searching for valuable information. A lot of the time they will find it from books, audios, videos, and events. The rest of the time the information they consume is worthless. But that will not stop them from finding valuable information that will make them a more valuable person, increase their influence, and increase their personal growth. By making themselves more valuable, they are helping their teammates to become more valuable too by influencing them every day.

When a low performing team player is watching videos on You Tube or listening to audios, it will very rarely be anything that will help increase their personal growth. They will most likely be listening to or watching something to do with music, or comedy. I am not against this, but to spend most of their free time doing this is a waste. You will also very rarely see a low performing team player reading a book on leadership or personal growth. They are not interested in searching for valuable information. They just rely on the information they received when they went to school or college. Educating yourself does not apply to a low performing team player. Even when they attend a training course in work, or a meeting, they are not present. When the training or meeting is finished, they will forget everything that was said or taught.

In 2009 when I started my self-education and self-development, I used to consume everything. I thought that everything I saw, listened to, or read would be helpful to me. It would have made me feel bad if I threw away notes, or only took in what I thought would make me a more valuable person. My book collection grew, and my notebooks grew. It wasn't until about 2011 that I realised I didn't need every single book I had, or every notebook I had. So, the books that didn't serve me I gave away to a used bookshop. The notebooks that didn't serve me I shredded. Over the years while constantly searching I will have missed some things. However, it will not stop me.

In 2009 when I started my first leadership position as project manager in Liverpool, I was reading the book *Start with Why* by Simon Sinek. If you have not read this book, then I highly recommend that you do. This was the first book I read where it felt that Simon was speaking to me. It was like he had written the book for me personally. There was one piece of advice that I took from Simon, and it was this, *"Leadership is not about being in charge, it is about taking care of the people in your charge."* In other words, it is your responsibility to take care of your team. It is your responsibility to help the team, to influence the team, and to lead the team.

I took my leadership and my personal growth a lot more seriously after I read Simon Sinek's book. I didn't consume everything within the book because some of the information did not make me a more valuable person. However, the valuable information I did consume is still very valuable to this day, and I have used this information when leading a team or helping an individual.

As you are reading this book, there will be some information that you will find more valuable to you. For other people it will be the same, but different information. You must take the valuable information you find and use that information to your advantage. That is what a highly effective team player does. They use the valuable information they must help increase their own influence and enhance their personal growth. But, most of all, they will use the valuable information they have to help their teammates. So, whenever you find valuable information, always remember to use it for yourself and for others.

A highly effective team player will come across valuable information every day. They are constantly searching. They are reading books when others are watching television. They are listening to personal growth and leadership audios when others are listening to music. They are watching personal growth and leadership videos when others are watching silly videos on You Tube. They are increasing their influence, increasing their personal growth, and helping teammates to do the same when others are standing back letting things happen. Highly effective team players make things happen. How much are you willing to do the same?

Make sure that what you see, listen to, and read is worth it. If it is not, then don't do it.

Understand others

When we talk about people skills, we usually mean leadership skills. What is the number one skill of any leader? Empathy.

A lot of people throughout the world do not know what it means to understand a person. I was one of these people for a very long time in my life. I didn't know how to really listen to a person, and I didn't know how to have empathy with a person. I thought that to empathise with someone meant that you must sympathise with them, but there is a difference. To sympathise with a person is when you have the same feelings as them, and you share them. To empathise with a person means that you don't have to share the same feelings as them, but you do understand the feelings of that person.

Having a discussion with a teammate and either agreeing or disagreeing with them is not the same as understanding them. A highly effective team player will always try to understand their teammate first before giving any opinion, feedback, disagreement, or agreement. However, there are team members who will think that when someone is understanding them, that they are agreeing with them. This is a mistake, so it's very important that when you want to understand one of your teammates first, ensure that you let them know that first. Otherwise, there will be a misunderstanding on your teammate's part.

A highly effective team player not only wants to communicate with his/her team, but they also want to connect with them. How does a highly effective team player do this? Firstly, they state very clearly to the teammate they want to discuss what their intentions are. They may say something along the lines of, "I want to understand how you visualise things, and why you visualise them in the way that you do. I am not going to give my opinion on the matter, nor am I going to agree or disagree." This sets the foundation for the communication. Then to connect with the teammate, they will listen more than they speak. They will use their ears a lot more than their mouth, and let the teammate do the talking. It's important that when you are listening, you understand the meaning of what the teammate is saying.

I talk a lot about influence, and how as highly effective team players and leaders that to increase our influence with others is how we lead. However, in the case of trying to understand our teammates, and connecting with them, we must allow ourselves to be influenced first. By understanding the teammate first before they can understand you, it is like you are passing the baton of influence onto them. They have taken the reigns of leader in this communication. What you are basically communicating through your actions of understanding is, "I am patient, and I want you to be understood first before you understand me. You are important to me." It doesn't seem like a lot, but doing this can be very powerful in communicating and connecting with your teammates. By passing the baton of influence, by the time you receive the baton back, your influence will have increased a lot more than you expected.

If you were to do the opposite and try to get your teammate to understand you first, then they will not listen to you. They will probably not be aware or know how to really understand someone, as you now do. So, because they think you don't understand them, they will try to talk over you to help you understand them, and the communication will ultimately breakdown. They will use their mouth a lot more than they will use their ears. But, if you do what we discussed above, and understand the teammate first, then they are a lot more prepared mentally to listen to you. They will want to understand you, and by doing this they are passing the baton of influence back to you. Your teammate will want you to influence them and are happy to be influenced by understanding you.

When a highly effective team player communicates and talks to their team, they want to be understood by them. If they are talking to someone who is not listening to them, then that is a complete waste of their time, so they will stop the conversation. When listening to their teammates so that they can understand, they reassure their teammate that they are listening. They do this by saying things like, "I can sense that this is a very emotional subject for you. Or "I can tell that you are excited and want to talk about it a lot more." Or "If you would rather talk about this another time, I am always available to listen."

Other ways that a highly effective team player assures their teammates that they understand is by repeating what their teammate said. If there is more people involved in the conversation, then by repeating and rephrasing what the teammate said will allow the other team members to understand too. This is a good trait to have, and it is very effective when the conversation is emotionally charged, or the topic is very important to the team. A highly effective team player may say things like, "Can I just repeat what you said then..." Or "I think I understand you correctly, is what you mean…" Or "I hear what you're saying, is this how you're feeling about it…"

A low performing team player will want to be understood first, so will put all their energy into talking more than listening. Most of the time, they are misunderstood because what they're saying is mostly negative. People switch off when listening to negativity, unless both people are negative. The negative team members who I have worked with in the past, every single one of them were the same. If what they were saying wasn't understood, or nobody was listening, they would all do the same thing. They would become very frustrated, so their voices would get louder, and louder. Some of them would become very animated and kick things or push things over.

It's interesting to listen to a low performing team player talking to another low performing team player. They are both looking to be understood first, so they are both talking at each other. They are not talking to each other. In the end, nobody is understood, and the communication breaks down. In the end, it is up to us when we talk, but it is up to others on whether they listen to you or not.

Being part of a team, no matter what industry you work in, for it to work then we must understand each other. As a highly effective team player, we must allow others to be understood before ourselves. If the communication is between 3, 4, or 5 people, then we must still allow everyone else to be understood before ourselves. That is what passing the baton of influence is, and what it means to connect with your teammates.

What we must also do as part of a group communication is to step in from time to time. If a person is speaking to allow others to understand them, then as a highly effective team player, we must allow the person to be heard. If someone talks over the person, then we must step in and tell that person to wait their turn to speak. It is not being rude on our part; it is being courteous towards our teammate.

Understanding our teammates first and allowing them to speak is a trait of a highly effective team player and leader. But, what about understanding ourselves? Before we understand others, we must first understand ourselves.

To do this we must continue working on ourselves every day. When working on ourselves we will understand our own strengths, weaknesses, likes, dislikes, and our capabilities. Highly effective team players work on themselves every day to understand themselves. I urge you to do the same.

To really listen to a person, what are you listening to? We listen to what they say, the tone of their voice, how loud they are speaking. But, to really listen, we must listen to how they feel, and how they think.

What do you focus on? Strengths or weaknesses, or both?

CHAPTER 8
MOTIVATION

Strengths and weaknesses

Most people throughout the world focus on weaknesses and neglect their strengths. I am here to tell you that focussing on weaknesses in competency is wrong. You must focus on strengths.

Throughout the world in most industries, it has been embedded into people (us included) to focus on the areas that we are weakest in, when it comes to our competencies. The idea is that if we focus on weaknesses, then we can try to improve those areas. However, while we are focussing on weakness, we are neglecting our strengths. We are not practicing them, so we are not fulfilling our competency potential in those areas of strength.

The same applies in business performance too. Businesses focus on the areas that they are not doing well in and try to improve those areas. By not focussing on the areas, they are doing well in, over time those areas will become weaker.

Highly effective team players want to fulfil their potential, so it is very important that they understand where to focus if they want to enhance their competencies. When I began focussing on my strengths instead of my weaknesses, I could feel a huge difference in my performance and so could others.

Highly effective leaders value the highly effective team players on their teams. One of the reasons is because they focus on their strengths and not their weaknesses. As you are working on yourself every day, it's time you start focussing on your strengths in competency. You are strong in these areas because you don't need a lot of effort to perform. However, your focus needs to be intense, not average. If you only give average focus, your strengths will remain average. If you give intense focus, then you will become an excellent, highly effective team player. Highly effective leaders are looking for excellence, and that will be you when you start to focus.

When it comes to weaknesses, these are the areas where you do not have the natural ability to perform without a lot of effort. Everybody has weaknesses. If you focus on your weaknesses instead of your strengths, then you won't become an excellent highly effective team player, and your team will not benefit. You will improve in weakness, but you will not be able to fulfil your full potential in your competencies. It doesn't matter if you give intense focus to your weaknesses, at best you will become average, not excellent. An average team member is appreciated, but they are not as valued as an excellent highly effective team player.

When it comes to a highly effective team player's character (not competency), they know that they need to focus on their weaknesses. Just like in competency, they need to give intense focus on their weaknesses, not average focus.

A highly effective team player knows to work on their competency and character every day by focussing on both their strengths and weaknesses. That is how a highly effective team player becomes excellent and will become the most valuable team player they can be. A highly effective team player realises that by improving their character in weakness, and improving their competency in strength is their secret to success.

As you are working on yourself every day to become an excellent highly effective team player, and work on being the most valuable team asset you can be, focussing on your strengths in competency and your weaknesses in character will be your secret to success too. Team members will hold you in very high regard, and leaders will value you to the highest level. The opportunities will also open to you very fast.

This book *"The Team Player"* has been written to help you to improve and develop your character. To increase your influence and become a highly effective team player, character is where you must put the most focus. Almost 90% of your results will be down to your character, not competency. You must not neglect your competency and work on your strengths to help develop them. Otherwise, you will not become the excellent highly effective team player that you are capable of being.

Your team deserves to have an excellent highly effective team player like you, so make it your responsibility to work on yourself every day to be just that. Your character will need to be as strong as possible if your team want to build a relationship with you. Your character will need to be as strong as possible if your team want to be influenced by you, work with you, and accept you for who you are. When you have built relationships with your team, and they want to be influenced by you, it will be your competency that will turn your influence, and relationships into actual results.

Character and competency go hand in hand. Your competency will enable you to take the teams, and the organisation's vision, and make it your own. It is your competency that will enable you to contribute to the team's and the organisation's mission.

When character and competency are working together in harmony that is when you will gain trust from your team, your leader, and your organisation.

If your character and competency do not work in harmony, then you are on the path to becoming a low performing team player. Team members will not want to work with you, they will not be influenced by you, and they will not really accept you. I have worked with many people who have a very high level of competency, but their character was at a very low level. It doesn't matter if you have a very high competency, with a low-level character, it will not work. You probably know people in your organisation, or even on your team who have high competency, and low character. What do you think of them?

Let's try and understand the reason why this happens so often, and why they are still a part of the team and the organisation.

On the teams I have worked with in the past, there have been many team members who had a low-level character, but we still worked with them. The reason we did was because there wasn't anybody else to do the task they were doing. They were good at the task, but their behaviour was unacceptable at times. However, we tolerated them almost out of necessity.

There was also more than one team member who had a low-level character, and they shared the same negative values. It was difficult for me because whenever I challenged one of them to improve their behaviour, they would team up against me. I found that to be very hard because I wasn't strong enough to keep challenging. I just accepted them for who they were, even though they were dragging the team down.

You will only ever see a highly effective team player working with a low performing team for one reason only. That reason is to turn the low performing team into a highly effective team. I have not seen this happen very often, because usually a highly effective team player will not work with a low performing team player. But, on the occasions where I have seen this, it was an amazing experience. To change a team from a low performing team into a highly effective team takes a very high-level character, and competency working in harmony. There is only a highly effective team player who can pull this off.

It was during my time as engineering technical manager in 2018 working in London, that I witnessed a low performing team be transformed into a highly effective team. I appointed a guy called Terry to join one of my technical teams. Terry was a very experienced and competent person. But most importantly, he had a very high-level character. It was his character that made my decision for me. Anyway, I advised him to help turn the technical team around and improve their performance. Within the first week, Terry showed the team what success looked like. He set his vision for the team and led the team from the front. After a month, the team's performance had improved nearly double. Attendance went up, projects were being completed on time, and every team member bought in to Terry's vision. I was so happy and excited that I wished I could have been part of that team and their journey. I helped them whenever they needed it, and it was a pleasure to lead them.

Terry is a highly effective team player and leader. I learned a lot from him, and it makes me smile whenever I talk about him.

To become a highly effective team player, we don't need more resources, we need more resourcefulness. We need to make the best of what we already have. Not wish for things and make excuses because of the things we don't have.

Have certainty

Whatever you do, never try to "wing it." If you do that you will lose your credibility, your integrity, and damage your reputation. Have the courage and have 100% certainty on what you don't know.

I have been part of the engineering industry since 1999, and in that time, I have worked with an automotive company, a gas and heating company, and many railway companies. All these types of companies have a lot of things in common, but one of the main things is "gossip." You probably know this yourself in the industry that you are working in. I don't think one day went past when I didn't hear a new rumour about someone or something, throughout my whole career. These rumours usually came from people who either just made them up, or they heard something and got the totally wrong end of the stick. It was incredible because as I climbed the leadership ladder, I found that no matter what level you are in the organisation, the rumours are spread everywhere. They could have come from the highest level in the organisation or from the lowest. Everybody spread the gossip, and everybody knew about it.

The reason people spread gossip or rumours was to try and get everybody else to see them as "go to" people for information. They thought that this would make them more credible within the organisation or their department. But most of the time the information they spread was either untrue, or completely off from the actual truth. So, most of the time, their credibility would be lost because they were spreading misinformation. I found it strange that people who didn't have a clue what they were talking about, would spread information that they were not even 50% sure of.

Rumours and gossip can come from every level of the organisation or industry, not just from the front-line teams. Sometimes it becomes a rumour by default because the CEO says something they believe to be true but doesn't 100% know it. Then someone who was in earshot of the CEO may spread what they say, and it turns out to be untrue. That will discredit the CEO because everyone will know it came from him/her.

Others may receive information from someone they confide in 100% and would not believe that they would give them false information. So, with that belief intact, the information will be spread, which again turns out to be untrue. This will discredit the person spreading the information, and the source of the information, which is the person who they 100% confide in.

A highly effective team player at all costs will not spread any information unless they know 100% that the information is a fact. They will never spread something that they are less than 100% sure of. A highly effective team player values very highly their credibility and integrity. If they share something that they are unsure of, they risk losing their credibility and integrity, so they will avoid doing that. If you spread something that you do not have *100% certainty* that it is true, then you are placing a very high risk on yourself. You will lose your credibility and integrity, and your reputation as a highly effective team player will be damaged.

When I started out my leadership career as a project manager, I learned very quickly that I needed to think before I spoke. But not only that, I needed to ensure I had certainty over what I was saying. A highly effective team player and leader will always have certainty over what they say, otherwise they won't say it. Having certainty is a trait of all highly effective team players and leaders.

When I was an engineering production manager in Scotland in 2011, I learned a very valuable lesson. Never try to "wing it". Whenever I was on dayshift, it was my job to present the morning brief to the senior leaders on the performance of the trains from the previous day, and what engineering issues had happened. Part of the briefing was to give details on whether the engineering issues had been fixed, and if the trains were serviceable again.

I remember one engineering issue had caused over 100 minutes delay in service (this happened at least once a week). Anything that caused over 100 minutes delay had to have a thorough investigation, and a report written. I was not fully informed on what happened, or if the issue had been fixed or not. So, when I was asked the question, I was so frightened because I didn't know, that I told my boss that the incident had been fixed. When in fact, my boss knew more than I did, and he was trying to find out how much I knew. Unfortunately for me, the issue had not been fixed and the train would be unserviceable for another day or two.

As you can imagine, my boss was not very impressed to say the least. He gave me a right good "telling off" in front of everybody in the room. I was so embarrassed because there were people in the room who I managed. To be honest, a "telling off" is quite light compared to what I really got from my boss.

So, from then on, I had learned to always be 100% certain, and if you don't know something, just be honest and say you don't know.

Another lesson I learned was to make sure that my team had 100% certainty too, whenever they were relaying information. I did not want them to make the same mistake as me by "winging it." So, whenever we had a team meeting and I asked a question like, "what happened when...?" Or "Do you know how this happened...?" If they gave me an answer which explained that they knew what happened, I would always ask a follow up question like, "How do you know?" I wasn't trying to catch them out, I was trying to help them be 100% certain.

There were occasions when certain team members tried to "wing it" with me, however I would always tell them about the mistake I made, and that they should not do the same. This helped them and enabled us as a team to ensure we always had 100% certainty.

Since making that mistake in 2011, I have become very comfortable with telling people "I don't know." For me, being able to do that is good leadership. I will never tell anyone that I know something when I don't. If you do that, it means you are an insecure person. However, if you are comfortable with saying you don't now, it shows the other members of your team, and the leaders that it is ok to say, "I don't know." So, it will benefit those around you, which is also a powerful message, and shows good leadership too. Another powerful message is telling your team that it is not ok to "wing it" or say you know something, when in fact you don't. Highly effective team players will never try to "wing it." If they do, then they will lose their credibility and integrity and damage their reputation.

A low performing team player on the other hand are insecure people, and it is highly likely that they will try and "wing it" when they don't know something. This brings out their insecurities to the forefront. A highly effective team player is a very secure person and will always say they don't know. The thought of "winging it" doesn't even enter a highly effective team player's head.

Throughout your organisation and your industry, no matter what you say, your credibility and integrity is at stake. So, ensure that you have 100% certainty on everything that you say, that includes when you say, "I don't know."

If you want to gain trust from your team, and for them to allow you to influence them, you must *have certainty*.

To have certainty is a very valuable lesson to learn and should be practiced every single day of your life, and every time you speak. You never know who is walking by and listening to what you're saying. It doesn't matter if you're in the lowest position in the organisation, or the CEO. Every single person must have certainty every day.

When working on ourselves, we must turn the lesson of having 100% certainty into a habit. We must practice this so much that it becomes second nature, and it must feel very uncomfortable to us if we ever try and "wing it."

We must also help others who we find that do try and "wing it." Sometimes people try to "wing it" out of fear, like I did in 2011. By helping others and showing them that it is ok to say, "I don't know" will show good leadership.

Don't let ego, pride, or even fear get in the way of learning the lesson to have certainty. The only person who will lose out is you.

Motivate don't manipulate

Being logical stimulates a person's mind. Being motivational stimulates a person's heart.

Have you ever been manipulated by someone within your team? Have you ever seen any of your teammates or colleagues be manipulated? I have been manipulated, and the bad thing is, I knew I was being manipulated. I just didn't have the courage at the time to challenge back and intercept the manipulation. It was not a nice feeling.

Manipulation is a very negative and selfish act. When you manipulate your teammates, you are negatively influencing them for your own personal needs. When you manipulate your teammates, it causes anger and resentment within the team. Your teammates will lose trust in you very quickly if you manipulate them.

Highly effective team players will never try to manipulate a teammate, or anyone else within their organisation for that matter. A highly effective team player will only ever try to motivate their team. By motivating your teammate, you are looking for a win/win situation for both. It is a very positive and selfless act. When you motivate your teammates, it makes the team happy, and they will take pleasure in being influenced by you. When you motivate your teammates, you are building trust within the team, and your influence will increase.

Whenever you have been motivated, it has come from something outside of yourself. Motivation is external. So, it requires somebody, or something to carry out an act that motivates you. For example, a motivational video on You Tube. You have been motivated by the video, e.g., the video is the external thing that has motivated you. When something external has caused you to take a certain action with your team, you are motivated. When something internal from within yourself has caused you to take a certain action with your team, you are inspired. Inspiration is internal and comes from within yourself.

Even though I have been manipulated by some of my line managers and teammates, I have also been motivated a lot more. I have worked with some highly effective team players, and leaders, and they were great at motivating me.

A highly effective team player knows how to motivate their team because, like me, they have been part of teams and worked with other highly effective team players who were able to motivate them too. So, they have a lot of experience in motivation. When you motivate a teammate, you enable them to take a certain action because of you. As a highly effective team player, the fact that you are part of your teammate's team will motivate them.

When I became a manager in 2009 and started to lead teams, I found it difficult to motivate the team initially. It felt like me being the manager wasn't making any difference, and that the team were doing everything the same as we were before. I wanted to make a difference, I wanted to help the team improve. The team I led were technicians and their job was to maintain trains. This is a very technical job, but the environment can be de-motivating. Especially when you are underneath a train getting filthy, and you see managers walking by you wearing a suit on the way up to their nice, clean, warm office. That was me when I took my first managers job. However, I realised that if I was to motivate my team, I needed to lead from the front and by example. So, one or two days a week, I would put my overalls on, and I would work with the team on the workshop floor. I would get underneath the trains doing inspections and trying to solve any problems with the team. I even went out into service to visit broken down trains. I wanted to motivate the team, and I thought the only way to do that was to be on the same level as the team. I didn't want to be some "ivory tower" line manager. I wanted to be a motivating leader. I wanted to be a highly effective team player.

I think it's very important that if you are a line manager, team leader, or a supervisor, you must also be a highly effective team player. It was my responsibility to lead the team, and results were my responsibility too. But it was also my responsibility to motivate the team, and the best way I believe to motivate a team is to lead from the front, and lead by example. By helping the team and working in the same dirty environment as them and seeing things from their level did motivate the team. The camaraderie between the team improved, we hung out socially a lot more, and we trusted each other. That is what a highly effective team player does.

One of the main tasks I worked on with the team was re-building the train bogies. The bogies are what hold the wheels and axles in place, along with other safety critical equipment. Our team had a bogie expert, so I took the lead from him. I could see that he really enjoyed leading his leader, and it motivated him a lot. The rest of the team fed off watching their leader being led, and they too took motivation from it. I was being led and influenced, but at the same time my own influence increased. Being part of the bogie building team and helping them where they needed it enabled us to improve and fix more trains in one week than we had before. All because we were highly motivated.

I must say I had a lot of fun being part of that team, and they had a lot of fun working with me too. Every Monday we had a team brief, and we would discuss the work we did the previous week. The team were excited to present their results because each week they were always improving. I made sure I sat out of the briefing and let the team run with it. I just watched and most importantly listened to them.

We set our own targets, and I vowed to work alongside the team on the workshop floor and "get my hands dirty" every week. I did this because I knew this was the best way to motivate my team, and most of all it was so much fun.

I've also been part of teams who were manipulated more than motivated. As I discussed earlier, I have been a victim to manipulation. When I was engineering technical manager in London, my line manager manipulated me very aggressively. He was using his position of authority trying to persuade me that I needed to fallout with my colleagues and blame them for things that were going wrong. Manipulation was the norm in that company, and I was oblivious to it even happening. So, I went along with my line manager, and I began to fallout with people and blame them when things went wrong.

In the end it backfired on me, and I started to get blamed for a lot of things. So, after a while, I almost broke down and I left the company.

Manipulation is a very selfish act, and when it happens to you it is highly embarrassing. It causes you to lose all your confidence in yourself, and any hope that you had.

Motivation is the complete opposite. It is a very selfless act, and when motivating another person, you are looking for mutual benefit and a win/win situation. It gives the other person more confidence, it fills them with hope, and most of all it makes the other person and yourself happy.

A highly effective team player will never use their position of authority to manipulate their team. They will use their position of authority to motivate the team, which will ultimately inspire the team.

Highly effective team players who don't have a position of authority can still motivate and influence their team. They can do this because of their character. They work on their character every single day to ensure it is the highest level it can be. Character is what brings out the best in everyone, especially highly effective team players.

Having the ability to motivate your teammates is a must for a highly effective team player. By working on your character every day, your influence will keep increasing and your ability to motivate will continue to improve.

Dictate from the back and people will run away from you. Lead from the front and your team will follow you.

Being an inspirational person is contagious. When you inspire someone, they want to become an inspiration too

CHAPTER 9
INSPIRATION

Have awareness

How do you like to do your work? Do you like to make it hard by being un-inspired, or do you like to make it easy by being inspired?

When a highly effective team player motivates their teammates, it is highly likely that those teammates will also be inspired. A highly effective team player can influence their team both when they are present with the team, and when they're not present with the team. Hence, they are both motivational and inspirational.

I was working as a consultant project manager for London Underground Tube Company in early 2019. My role was to manage the overhaul of the fleet of trains that ran on the District Line. The team I had consisted of 9 consultant engineers who had never worked together before, or even met each other before. The role of the engineers was to validate all the overhaul instructions, vehicle maintenance instructions, and bill of materials, tooling, and facilities. It was a huge project that was planned to last a year. The validations that our team were required to complete was huge, so I had to bring the team together and motivate them. Otherwise, the project wouldn't have lasted a year, and our consultancy wouldn't be offered any further projects. There was a lot at stake for us to pull off.

It took about a month to bring all the 9 engineers onto the project, and then a couple of weeks to delegate the first phase tasks to each team member. 4 of the engineers were contractors, and the other 5 worked for the same consultancy I did. So, at first there was a little bit of "us and them" between the contractors and the consultants. This made my position even harder, because not only did we not know each other, there were also people who didn't like each other right from the start.

The goal I set for the team was to complete our tasks on time to the best of our ability. But the real goal I set was for us to get a long and enjoy working together. Not everyone bought into that goal right from the beginning, so I arranged a social get together to try and get to know each other first before we really got stuck into the work.

We went paint balling together, and then we went for dinner and drinks later in the evening. It was a good day and we all got along great. We talked about our previous jobs, our families, friends, told stories, but most of all we were happy to be with each other from then on. It was a huge success.

We went back to work on the Monday morning with a new attitude following our successful social encounter. We were ready to work as a team and help each other.

The first week of starting the project for real, I did a lot of observing. I wanted to see if the team were relying on me or relying on each other. I advised the team that I would not be a micro-manager or be on their back each day looking for updates. They had the autonomy to make their own decisions and work with the client as they saw fit. I tried to ask a lot of open questions that would enable them to think for themselves and come to their own answers. By asking questions, I was not telling them what to do, they were "telling themselves", what to do. I found it interesting because they didn't realise that's what they were doing.

At the end of the first week, the team worked well together, and without any of my input or direction they had already drew up a plan for the second week. Mark and Dean were two members who worked for the same consultancy as me, and they took me through the plan. I could tell by their voices that the plan excited them, and I could tell by how they came across that they were both leaders. I saw them as being the leaders when I wasn't there.

I didn't tell Mark and Dean that I wanted them to be the leaders, I wanted to see them lead naturally. So, in week 2, I did a lot of observing again. This time I wasn't observing if the team got a long because they obviously were. I wanted to see who the leaders were. At first Mark and Dean took charge, and because of their leadership they inspired the rest of the team to lead too. It was amazing to witness a team building themselves into a team of highly effective team players. Even though I was responsible for this team, it was beginning to feel like I didn't even need to be there anymore. Which is a great feeling.

I had set the foundations and the goal that we should be a team who got along and worked well together. Now this was happening, I just needed to assist the team whenever they needed me. Mark and Dean were learning every day, and every day they were inspired from within because they were making things happen on the project, without needing me to motivate them. The rest of the team were inspired too because they were also making things happen. All I had to do was check on progress with them every Friday, and that is when they took me through the following week's plan.

Every week when I checked on the team's progress, it was great to see that everything was on track, and they were even ahead of schedule on some tasks. During my one to ones with the team, you could tell in their voices that they were enjoying what they were doing. You could really feel that this was a team, and the team was getting stronger every day. Everyone was praising each other and helping whenever they were needed.

I remember one day I walked into the train depot, and three members of the team were witnessing the strip and rebuild of a train gearbox. I observed them from a distance, and I could see that they were working in harmony together, and they also had two members from the client team in harmony too. When they finished, Mark came over to me and told me about all the achievements for the week so far. I was so happy for them, and I couldn't stop smiling for the rest of the day.

I felt that I was part of the team, but really, I was in the background. But that was fine with me because I wanted the team to be the highly effective team players. I wanted them to be the highly effective leaders too. All I did was set the tone and the vision for the team, they then took that vision and made it their own. They were inspired and had learned how to become inspirational and inspire others.

Over the whole year throughout the project, the team kept impressing me every day. Especially Mark and Dean, as they continued to motivate the team and inspire them to be the same as them. They led from the front and led by example. They stayed late when they needed to. They got in early when they needed to. They even worked weekends for no extra pay. That is how motivated and inspired Mark and Dean, and the team were. I couldn't have thanked them enough.

Mark and Dean were highly effective team players, as were the rest of the team. But Mark and Dean kept inspiring, they kept the team together without me needing to be there. The team loved working with them and would go the extra mile for them. It was amazing to witness.

I remember waking up every day when I was part of this project excited to go to work. I couldn't wait to see what the team were going to achieve each day. It was like going to a theatre every day. I would go to work and just observe the team at work. I would grab a coffee with them at break times, and then when they went back to work, I would watch them. It was one of the most pleasurable jobs I had ever had.

When you know how to motivate a person, you eventually learn how to inspire a person. When you know how to inspire a person, your influence will increase that much that you don't even have to be present or in the same room. That is what inspiring a person is.

"What are you going to do differently?" This is an overused phrase in the world of work. Most of the time nobody does anything differently. To do something differently, you must do something differently.

What are you going to do differently?

Have awareness

If you don't have awareness, then you don't know that you don't have awareness. If you can't see what you need to have awareness of, then you don't know that you can't see what you need to have awareness of. To grow to your full potential, you must have awareness. How do you raise your awareness?

By becoming a highly effective team player there are a lot of things that you need to be aware of, and one of those things is to have awareness. When it comes to the differences between a low performing team player and a highly effective team player, one of those differences is how aware they are. A highly effective team player's awareness level must be higher.

To help you raise your level of awareness, you must keep working on yourself every day in both the competency and character areas. Reading this book *"The Team Player"* is a great start in making you more aware. However, there are hundreds of self-development books out there that can raise your awareness level even more. So, when you finish this book, don't stop there. Keep searching for more books and materials that can help you continuously improve as a highly effective team player, and eventually a highly effective leader. You will learn about more people (besides me), and their stories throughout their lives and career. The more stories of highly effective team players and leaders you can relate to, the better it is for you.

If you are enjoying this book so far, and you feel that it is making you more aware, then congratulations to you and keep reading. To help make you even more aware, increase your influence, and help you raise your leadership level then learn about my other books that I have published. You can check them out at https://highlyeffectiveleader.com or you can find them at www.amazon.com.

I have been reading self-development and leadership books since 2009, and with every new book I read I am becoming a lot more aware. I read daily, and this is a part of how I work on myself every day. I encourage you to do the same. We discussed earlier in this book about strengths and weaknesses. To work on your strengths in competency, reading will help with that. To work on your weaknesses in character, reading will help with that too. Being deliberate in these areas is a must…every day.

Highly effective team players take action every day with their development and personal growth. They are also constantly seeking response and feedback from their leaders and teammates on how they are doing. Highly effective team players have very high levels of self-awareness, but there are times when they can't see everything. So, feedback from teammates helps them in the areas they can't see for themselves.

Being aware and being focussed is the goal when it comes to a highly effective team player and their team. One thing a highly effective team player has great ability in is the ability to focus on what the team is trying to accomplish, the mission. However, whilst being focussed on the mission, they also have awareness of what the organisation is trying to accomplish, the objective. By being able to do both, nothing will be lost. The team will be able to achieve both their own mission and still contribute to the organisation's objective.

We have talked in detail about motivation and inspiration. As a highly effective team player is working on themselves every day, and increasing their awareness level every day, they are in a great position to help their teammates increase their awareness level every day too. As we now know, a highly effective team player can motivate their teammates with ease. A highly effective team player is inspired from within to help develop their teammates and their personal growth. They don't need any motivation from anyone else to be able to do that. A highly effective team player's mission is to grow their teammates and increase their influence.

This is who a highly effective team player is. They are part of the team so they can add value to the team. It is what they do. They are so aware of themselves, and know themselves inside out, better than anyone else. They are fully aware of their own purpose, and they are fully aware of their own passion. They use their purpose and passion to help the team and give them the most benefit possible from it. A highly effective team player is not seeking any type of reward or recognition for it, they are happy to be in the background remaining highly effective.

For many years as I was going through my career, and living in different places, I caught myself looking backwards a lot. I would think of Scotland and wonder "I wish I was still in Edinburgh." Or, I would think of my hometown Liverpool and say things like, "I should have stayed home with my family, and I miss them." These are nice things to think about, but the past is something we shouldn't dwell on or criticise ourselves about. When we learn something new, we imagine what our past life, or our past job would have been like if we knew this new thing we have just learned. We think, "I should have done this if I'd known." Or "I could have done that if I'd known." The past is something we should think about and reflect on, not dwell on. What we should focus on is the future.

Now that we are learning new things every day because we are working on ourselves every day, we should look forward. We should be excited for what's in store for us with our newly found skills and insights. It is thinking like this that will keep our awareness level high, and it is thinking like this that will raise our awareness level every day.

When I stopped dwelling on my own past and instead began to reflect on it, I started to learn more about myself. I reflected on my past mistakes, learned the lesson from those mistakes and vowed to myself to do my best never to make those mistakes again. The more I reflected, the more I learned. The more I learned, the more aware I was becoming. However, I wasn't just becoming more aware of myself, I was becoming more aware of those around me. It's amazing what you feel and realise when you change how you think about your past, and use it to learn, not to beat yourself up or criticise yourself.

When you look inside yourself right now, how do you feel? Are you beating yourself up? Are you criticising yourself? A highly effective team player is constantly looking inside themselves and learning about themselves. They are also looking at their teammates and learning about them too. A highly effective team player is so aware of both them and their teammates. By being aware of others on your team, you will know exactly how to motivate them and how to influence them.

The more you learn about yourself, and you start doing things differently, you will not only become a highly effective team player, but you will also become a better person overall. Highly effective team players don't just have to do things better, they must be better. Raising your awareness level isn't just about your competency and what you can do, it is about your character, and who you are. We discussed that to be a highly effective team player that competency and character must work in harmony. Well, to raise your awareness level, do better, and be better, then again competency and character must work in harmony.

How do you feel about your teammates, your friends, and your family? If a person is not important to you, then how can you make them feel important?

Ignore negativity

If you value negativity, then your life will ultimately be a negative experience. If you value positivity, then your life will ultimately be an amazing experience. It's your choice what you value.

Every person on your team and within your organisation has values. The low performing team players will have values, the average team players will have values, and you as a highly effective team player will also have values. Everyone's values will be different.

Teammates who are either average or low performing team players do not have the same options as a highly effective team player. It is your values and your character that raise you above the rest of your teammates as a highly effective team player and provide you with more options. You will use these options to enable you to be more in control of your future, and most importantly yourself. As your character level keeps increasing, so will your influence, and your teammates who have the right attitude will choose to follow you. People with values that are like yours will choose to follow you too.

However, there are low performing team players in many teams also, and their values are usually the opposite of a highly effective team player. Low performing team players have a much lower character level than a highly effective team player. They have the objective to bring the rest of the team down to their level. They will not try to raise their character level like a highly effective team player would because that will conflict with their values. They want to poison the team, and in my experience, they usually spread their poison successfully.

Early on in my career, I believed that I had to be like the low performing team players because it was uncool to become highly effective. I saw it as being a "goody two-shoes," and there was no way I could be seen to be like that. The reason I believed this is because the other, more experienced low performing team players were poisoning my mind. If anyone on the team was trying to raise their character level or increase their influence by becoming more effective, then we would pull them right back down to our low performing level. We couldn't let anyone become the "main man/woman" or the "superstar" of the team. Unfortunately, it was this thinking that made us one of the lowest performing teams in the organisation. Luckily for me I was only 19 years old, so I had a lot to learn.

It wasn't until I was 22 years old that I started to "see the light" and realised that I could make a difference. I had the opportunity to become a highly effective team player, and eventually a highly effective leader. I changed my attitude from being negative and "cool," to being positive and "ambitious." My relationships with others within the organisation got a lot better, and became positive, especially with the other highly effective team players. Then I began to build positive relationships with my own line manager, the other engineering managers, and eventually the engineering director. The low performing team players on my team did not like this at all. I got teased by them, and was being accused of "boot licking," and "brown nosing" to the bosses. It got a lot worse than this too, in fact you can call it bullying.

It was up to me to decide if I was going to listen to the low performing team players/bullies, and allow their insults to get to me, and then lower myself back down to their level. Or I could decide to keep raising my character level and work on myself to become a highly effective team player. I chose to do the latter. Low performing team players who spread their negativity across the team have already made their decision to behave in that way. They have no ambition to raise their character level or increase their influence in a positive manner, they are committed to increasing their influence in a negative manner. Their ambition is to try and shame any team member who wants to make a positive difference and bully them until they lower themselves to their low performing level. They use manipulation tactics, and when they succeed, the whole team begins to suffer.

Highly effective team players know how to handle working with low performing, poisonous team players. They can control themselves to not be influenced by their negativity and will never drop their high-level character to anywhere even close to a low performing team player. They interact with low performing team players only when they need to, and they will end any relationship with them when they need to. Highly effective team players have the strength and awareness to know when a relationship is getting worse, and they can terminate it when needed. They can also do this if their boss is low performing and poisonous too. Or, if the organisation they work for has a poisonous culture, then they will know when to end the relationship and move on.

Having the courage and strength to end poisonous relationships with teammates, leaders/bosses, or even your employer is imperative as a highly effective team player. Having self – control is a very important trait of a highly effective team player, and they know exactly how to stop anybody trying to control them.

Building relationships and influencing the teammates who are vulnerable to being controlled by low performing team players is a responsibility of a highly effective team player. They want to help their teammates in raising their own character level and increasing their influence in a positive manner. They do not want their teammates being caught in the clutches of poisonous team members. Highly effective team players want to create other highly effective team players. Whereas low performing, poisonous team players want to create other low performing, poisonous team players.

One of my greatest achievements was building the strength and the courage to stand up against low performing, poisonous team players. It took me a while to build that strength and courage, and it was extremely difficult to stand against this negativity, but I learned how to do it and eventually broke free. There are a lot of good people I have worked with in the past who could not break free of the negativity, and they are still in the same place today. People find it especially more difficult when they don't have any highly effective team players who can help them, and who they can follow. In the end, people give in to the negativity, and let themselves be controlled because it is the easier option.

It's the easy option to remain in a negative team and keep negative relationships with teammates. The ironic thing is, the negative teammates want things to improve, but they focus on the wrong things. The expect everything to change around them. However, if they want things to improve, they must first improve themselves, not everything else. So, because they are unwilling to change themselves and their attitudes, everything remains the same.

The purpose of writing this book *"Be a Highly Effective Team Player"* is to help you to learn how to raise your character level, increase your influence, and spread positivity through your team and organisation. As you do this, you must ***IGNORE NEGATIVITY***, and you must avoid the low performing, poisonous team players. You must also be committed to your teammates who need your help, so that they can raise their character levels, and increase their influence just like you.

As I started to climb higher on my journey, and the more people I helped along the way, I received quite a lot of praise from teammates and colleagues. I also found that from people who I considered to be friends once, were not very supportive at all. They would be trying to bring me down to their level, but I didn't let them, and I ended our relationship. This is something you must do too if you have to that is. It is not easy, but it is the only way you can continue to raise your character level and increase your influence.

You must not let anybody hold you back, whether it is a member of your team, a friend, a family member, or even your boss. Stay true to yourself and your journey, and the right people will notice you, and the right people will support you.

When we build great relationships, we build trust. With the support and trust of our great relationships, we will become stronger and build the courage to stand up against negative, low performing, and poisonous people.

To develop a strategy, you need a vision. To have vision, you need to have optimism

CHAPTER 10
PERSEVERANCE

Have optimism

People who are pessimistic see failure as final. They put failure down to an obstacle that is impossible to move or change. People who have optimism see failure as an opportunity to change or remove an obstacle so they can turn the failure into a success.

People who have optimism know that it's important to have a vision. They have an optimistic view and belief for the future. A view that the future is bright. A view that the organisation they work for is going to get better. A view that their department is going to get better. A view that the team they are part of is going to get better. A view that their teammates as individuals are going to get better. Most of all, they have a vision that they as an individual is going to work on themselves every day to get better.

Highly effective team players know that to turn their vision a reality, then they need to take action. But what exactly do they take action on, and how do they take action? Highly effective team players take action on their team. They lead from the front, and firstly show their team the vision, then take them on the journey to turn that vision into a reality.

Highly effective team players have optimism, and before you can have a vision, you must first have optimism. When you have optimism, it will help you to raise yours and your team's level. However, you cannot get to the next level on optimism alone. You must take action to bring your vision to life and make it a reality. By focussing on your vision, and taking action, you will become a highly effective team player. You will also be on the right path to building a highly effective team.

As competency and character work in harmony, so too does optimism and vision. To have optimism, you must first have a vision. To have a vision, you must first have optimism. When your vision becomes 100% clear and complete, your optimism level will increase. As your optimism increases, the urge to take action grows. As you continue to take action, the more your vision will become reality.

To make your vision a reality, you will be required to make sacrifices for the sake of that vision. When you have optimism for your vision, you will realise that making sacrifices is necessary. However, to begin making sacrifices and then continue making them, you must become disciplined. By having the discipline to make a sacrifice, your vision will become the priority until it becomes your reality.

Will sharing this book, or sharing what you are learning from this book help your teammates? If the answer is yes, then that is because the vision you have for your team is to help them become highly effective team players just like you. So, the first sacrifice you are making is reading this book, which takes up your time. Then the second sacrifice will be sharing what you are learning, which again takes up your time. But these sacrifices are necessary for you to help turn your vision into a reality. The action you would be taking is reading the book, and then sharing it. Hopefully you are starting to see the correlation between sacrifice and taking action, and how important it is for both to happen.

Every vision has a price, and that price is sacrifice. If you have a vision to become a highly effective team player, then you must make the sacrifices necessary and have the discipline required to work on yourself every day and turn that vision a reality. If you have a vision to help your teammates become highly effective team players, and build a highly effective team, then you must make the sacrifice of your time. You must also have the discipline to meet with your line manager, if he/she has final say on what happens with the team. If you have the autonomy to make things happen, then you must make the sacrifice of your time, and have the discipline to meet with your team. During these meetings you must lay out your vision so everybody can see it with you. Then map out the action steps to turn your vision into a reality for both you and your team. Those steps could be taking your team through each chapter of this book, so you can share what you have learned on becoming a highly effective team player. If you have other action steps in mind, then they must be mapped out and acted upon. The price must be paid to turn your vision a reality for both you and your team.

Highly effective team players are ambitious, and they want to continuously improve. They also want to help others to improve too. To accomplish these things, you must be able to see what it looks like when it is accomplished (your vision). Your vision will create an emotion inside you that will not let you detour away from it (optimism). Your optimism will give you the strength to pay the price to turn your vision into a reality (sacrifice). Your sacrifice will give you the courage to not give up on your vision (discipline). Your discipline will enable you to accomplish what you saw in the beginning (result).

It is not easy to put into place everything that is required to make a vision a reality. As you climb the leadership ladder and take up higher positions throughout your career, more people will begin to rely on you. You may at times want to give up on your vision, but I am here to tell you, ***DON'T GIVE UP***. Your vision will benefit so many people, including yourself. The higher you climb; your influence will increase across so many more teams. If you continue to have optimism, your vision will become a reality, and a benefit for everyone.

As I climbed higher up the leadership ladder, my level of vision raised higher too. I always saw a vision of creating highly effective team players, and highly effective teams. But, as I started to take on more responsibility and more people, the harder it became to get the buy-in from everyone. Especially when these people were low performing team players. I had the optimism that I could influence these people, but on some occasions, I had to leave them behind. I left them because I couldn't have them dragging me, or the other optimistic team members who saw the same vision behind with them. We had to continue to move forward.

I got help from my team of optimistic team players who were on the journey to becoming highly effective. They tried to sell my vision to the low performing team players, but because of how stubborn they were, and the values they had, no matter what we tried they would not buy-in. I know for a fact that they could see the team changing. We were disciplined to persevere with making the sacrifices and taking the action required to make our vision a reality.

It gives me hope that some of the low performing team players did eventually change their ways and came around to the vision we had. Some of them even helped us in putting things in place and took the same actions that we did.

With optimism, vision, sacrifice, discipline, perseverance, and hope, you can make things happen and bring people along with you. For the sake of yourself, your vision, and your team, just **DON'T GIVE UP**. Your vision must become a commitment, if you commit then you are more likely to not give up on it. If you think you can turn your vision into a reality, then you will.

You are reading this book, so I believe that you have a vision, and I believe you have what it takes to make your vision a reality.

Successful people see failures as a steppingstone to ultimately achieve their greatest victories.

Live your values

By living your values, and leading from the front, you are building a stronger team, and a stronger culture. By sharing your values, you are creating highly effective team players.

What do you value, and how do you think? Everybody has values, and everybody has a way of thinking. Being part of a team of people were each team member thinks differently and has different values would be a very hard team to build. For example, a highly effective team player thinks very differently to a low performing team player. They also achieve very different results because of what they value.

So, to build this kind of team, it takes discipline, sacrifice, responsibility, and engagement. A highly effective team player will value these traits. However, a low performing team player won't. I've met and worked with a lot of people who were part of difficult teams, but they did not have the discipline, they did not make the sacrifices, they did not take the responsibility, and they did not engage enough to help make their team better. The reason for this was because they did not share the same values.

I have also met and worked with people who were part of difficult teams, and they did have the discipline, they did make the sacrifices, they did take on the responsibility, and they did engage enough to make the team better. In which they eventually did help their team improve. I was part of some of these teams too. We have discussed previously that to make something happen then we must be willing to work on ourselves every day and help others at the same time. That is the price we must pay.

We must pay the price and have the courage to increase our influence with our team if we want to see it improve and achieve better results.

For our influence to increase within our team every day, then we must have the respect of our teammates. I have a book called *"Manager to Leader"* and I discuss the difference between demanding respect from your team and earning respect. People who demand respect have false influence, and they are usually the boss. Teams will respect the position of the boss, but they will never respect the person if they demand respect. People who earn respect will be able to increase their influence authentically, and their team will respect them as a person.

We hear leaders talking a lot about culture, and how we need to change it. What we value, how we behave, and how we feel to be part of the team is what culture is. So, to change the culture, firstly we need to know what each team member values, how those values are making them behave, and we need to learn how they feel at this present moment in time.

Why do we need to know these things? Because to change the culture, we need to change the values of the team, so that we can change the behaviour of the team, and then help our teammates to feel differently about being part of the team. We need to help our teammates to feel great about being part of our team.

To be a highly effective team player we need to live and breathe the values that a highly effective team player believes in. We need to set the example and model those values and behaviours every day. This is the only way we can influence our teammates to change their values and adopt the same as ourselves. We cannot force them to change their values, it all comes down to our ability to influence, and how we set the example.

When it comes to determining your future, and what your vision is for your career, and your life, it starts with how you think. What you value in your life produces your thoughts, and then how you feel because of your thoughts. From your thoughts, you then decide what you are going to do, or what action you are going to take. From the action you then take will determine what results you produce. From the results you have produced will lead to the circumstances you are currently living in.

I talk a lot about working on ourselves every single day to become highly effective team players, highly effective leaders, and to continually grow. If you don't value working on yourself every day, then you will not become a highly effective team player, leader, and you will not continually grow each day. If you do value working on yourself every day, then you will think about working on yourself every day too. What you value is the beginning of all your thoughts, produces your first thoughts of the day, and all of your thoughts throughout the day. So, to become a highly effective team player, you must share the values of a highly effective team player, which will then fuel your thoughts of becoming a highly effective team player.

Whatever thoughts you have will then determine how you feel. If you currently don't feel like a highly effective team player, it is because you are not thinking like a highly effective team player. Your feelings influence what decisions you make and what action you take. If you want to start feeling like a highly effective team player, then you need to change your thoughts, and think like a highly effective team player thinks. When you start thinking differently, you will feel different.

How you feel will then influence what decisions you make, and what action you take. We can all be influenced by different people. So, what we are thinking, and how we are feeling can be influenced by different people too. If we are influenced to think and feel negatively, then we may be influenced to act in a negative manner. If we are influenced to think and feel positively, then it is highly likely that we will act positively because that will make us feel good.

The results we produce will come from the action we have taken. If we act in a negative manner, then we will produce negative results. If we act in a positive manner, then we will produce positive results. Whatever decision we make will determine what action we take. So, if we want to produce good positive results, then we must make good and positive decisions.

Our current circumstances have come from the results we have produced throughout our life and career. If we are not happy with our current circumstances, then we must go back to the beginning and change what we value. The root cause of our circumstances has come from what we value. A highly effective team player knows this, and to help others change their circumstances, they must first help to influence them to change what they value.

Ultimately, it should be the leader of the team to influence the team members to change their values and change their behaviours. But, if the leader is not capable, and isn't a highly effective leader, then someone else will have to do it. Otherwise, nobody will do it, and the team will not be a team at all. They will be a group of individuals, all with different values.

What usually happens with a group of individuals with different values is, they must follow certain rules to keep everyone in line. For example, what time they start work, who does certain tasks, when they do those tasks, and then what time they finish. If everybody shared values, and were striving towards the same vision and goal, then there would be no need for rules.

If the leader of the team is weak, then a highly effective team player will put their hand up to take over. A highly effective team will talk to each team member individually to find out what their values are. If everyone has different values, then the highly effective team player will share their vision and values with the team and influence them to adopt the same vision and values. Leading by example and showing the team the way is how they influence the team. Are you willing to take over from a weak leader, share your values, and lead from the front as a highly effective team player?

What is it you aspire to be? Do you aspire to become a highly effective team player? It's easy to aspire, it's very hard to achieve.

Do not quit

To become successful, you don't need to have the best resources, you need to be the most resourceful you can be. What you believe in, what you value, and who you are will push you to success.

Being part of a team and being a highly effective team player, you have a responsibility to keep improving. Being the best on the team is not what is important. The most important thing is to work on yourself every day, and to help your teammates to do the same. Your team are relying on you to keep working on yourself. They are relying on you to help them. They are relying on you to ***NOT QUIT***. Keep striving to reach the next level and even further and keep striving for your team to reach the next level and even further too.

So, ***DO NOT QUIT***.

At this moment in time, you and I have still not reached anywhere near close to our full potential. Your teammates have not reached anywhere near their full potential either. We all still have a long way to go to reach this level, so it is so important that we persevere and don't get disheartened and quit.

Try not to think of your teammates, or others within your organisation and industry as competition. People who are not at the same level as you are not your competition. People who are around the same level as you are not your competition. People who are ahead, and even way ahead of you are not your competition either. There is only one person who is your competition: **YOU**.

You are the only person you should be in competition with.

Highly effective team players are only in competition with themselves. They compare themselves to the person they were yesterday, and so should you. They reflect on the person they are today, and so should you. They compare themselves and look forward to the person they can become tomorrow, and so should you. If becoming a highly effective team player is a commitment, then you are the only competition you will ever have. So, do not quit on yourself.

By working on yourself every day to become better will keep you fully aware of how you are doing, especially when comparing yourself to the person you were yesterday. If you don't work on yourself every day, then you have no comparison because you will still be in the same position. For example, I read leadership books, or self-development books every day, mainly to improve my character. I do this to expand my knowledge and make myself better than yesterday, so the comparison is obvious, and I can be honest with myself. Working on myself every day also gives me the knowledge of how far I have come in a certain amount of time, and how far is left for me to come close to my potential. This is part of reflecting on who and where you are in the present day.

When thinking about and comparing yourself to the person who you want to and could become tomorrow, when you eventually reach your potential can sometimes be difficult. However, it is necessary to make this comparison. You will realise how far you have left to go to reach your full potential, and how much you need to improve. You will give yourself a reality check that we all need to have, and again be honest with ourselves.

When you are making these comparisons, make a promise to yourself. The promise is to not let failure, or the fear of failure to pull you backwards, or worse…quit. Instead use failure to learn from any mistakes you made yesterday, or today. When people talk about failure, they are not actually referring to failure as final, they are referring to learning something new about themselves or someone else. In your case you will be learning about yourself. When you are learning to drive, during the lessons your instructor doesn't say you are failing to drive when you make a mistake. No, he/she tells you that you will learn from that mistake. We should be the same way with ourselves when we make any mistakes. We are not failing, we are learning.

This trait is also an excellent trait to teach your teammates to make the same promise and commitment to themselves. The learning to drive analogy works well too because most will not give up on their driving as they want to pass their test and buy their own car. We must not give up on ourselves when working on ourselves every day and striving to become highly effective team players.

When you or your team are starting something new, it is essential that you must get comfortable with making mistakes. If you don't make mistakes at the beginning then you are not going to learn or figure out how to do things better. Some people call mistakes failures, and some people call mistakes… mistakes. It is up to you what you call yours, but the point is that we need to use our mistakes in a positive way to move forward. The most successful people and the most successful teams in the world have made many, many, many mistakes. The difference between them and the unsuccessful ones is how they learned from those mistakes, and then took action to move forward. A highly effective team player has made lots of mistakes too on their journey, and they know how to be comfortable with making them, and then learn from them. Mistakes will never make a highly effective team player give up on themselves, or their team.

As you strive towards reaching your own potential, and as you strive towards helping the team reach theirs, you must do it with high intensity. By keeping the high intensity up, it will energise you and the team, and your willingness to succeed will ever increase.

Whatever influence you have on yourself and your team today, it can always be increased tomorrow. If you think that you are great already, then you can always be greater tomorrow, make no mistake about that.

The question is: Do you have what it takes to increase your influence, become a highly effective team player, and strive to reach your full potential?

If you do have what it takes, then you must work on yourself every day, and most importantly not give up on yourself. Your full potential is in the distance, but it can be reached. However, it cannot be reached if you only work on yourself once a week, or even three times a week. It can only be reached if you commit to working on yourself every day.

You can only become a highly effective team player if you work on yourself with high intensity. You can only become a highly effective team player if you help others to work on themselves every day too and strive to reach their full potential. You can only become a highly effective team player if you remain positive, and work to improve your character every single day. You can only become a highly effective team player if you commit to becoming a highly effective team player. The only way to become a highly effective team player is:

DO NOT QUIT!

Quitting on yourself and your team is the easy way out. If you start something, why not work on yourself every day, help others, and make it a success that you can be truly proud of.

It would be great to hear your thoughts on how this book has helped you as you climb the leadership ladder. I would also love to know how you are doing by helping others to climb the leadership ladder, and beyond.

So, please feel free to email me and share your thoughts at:

tom@highlyeffectiveleader.com

To order my books, online courses and my other resources, please visit https://highlyeffectiveleader.com or www.amazon.com

It is ok to make mistakes, it is ok to fail. Failing does not make you a failure. Quitting makes you a failure.

ABOUT THE AUTHOR

There are hundreds of books that teach us how to be a "leader". Most of them are good and teach the right things. But, Tom Lawrence has seen very few that teach us how to influence our people, or take our personal growth seriously, or even lead if you're not in a "leadership position".

Tom wants to change that by writing books for people who are interested in leading and inspiring others.

He wants to help people who are not in a leadership position but are aspiring to become a leader, current managers who want to take the next step and become highly effective leader, and to help senior leaders enhance their leadership skills.

Tom began his career in 1999 as an apprentice mechanical engineer, working for an automotive company in Liverpool, UK.

After completing his apprenticeship he was made redundant from that company and had to find other work, that is when he joined the rail industry working for Merseyside's train operator. He worked there for six years, and during that time he achieved his degree in mechanical engineering, his master's degree in maintenance engineering, and he started his first leadership role as a project manager in 2009.

After leaving Liverpool in 2011, he has worked in Edinburgh, Glasgow and London where he currently lives. Tom became a chartered engineer in 2013, and is a mentor for new and upcoming engineers working towards their chartership. Working in these different cities he had remained within the rail industry and led different types of engineering teams, learning and practicing different leadership styles.

Tom has vowed to make it his life's work to help you become successful. He hopes the information that is provided through his books, goes some way towards helping you achieve all you ever wanted from your career as a leader.

Tom's Leadership Mantra

Learn to Lead, Grow and Increase your Influence

Tom's Why

To develop leaders and aspiring leaders into highly effective leaders so that, they can develop leaders and aspiring leaders into highly effective leaders.

Tom's Vision

Tom can see a world in which our leaders create environments that focusses on people and helps them to learn to lead, grow, and increase their influence for the good of us.